Managing Technology

Amy Zuckerman

- Fast track route to managing technology

- Covers the key areas of technology management from knowing which technology trends count and managing technology change, to devising the right level of technology training and and selecting technologies that match your business strategy.

- Examples and lessons from some of the world's most successful businesses, including ACUNIA, Nextel, and Seven-Eleven Japan and ideas from the smartest thinkers, including Bill Gates, Steve Jobs, Marc Adreesen and Timothy Berners-Lee

- Includes a glossary of key concepts and a comprehensive resources guide

OPERATIONS

06.08

>>EXPRESS EXEC.COM<<
essential management thinking at your fingertips

First published 2002 by
Capstone Publishing (a Wiley company)
8 Newtec Place
Magdalen Road
Oxford OX4 1RE
United Kingdom
http://www.capstoneideas.com

```
HD
45
.Z82
2002
```

CIP catalogue records for this book are available from the British Library and the
US Library of Congress

ISBN 1-84112-227-0

Printed and bound in Great Britain

This book is printed on acid-free paper

Substantial discounts on bulk quantities of Capstone books are available
to corporations, professional associations and other organizations. Please
contact Capstone for more details on +44 (0)1865 798 623 or (fax) +44
(0)1865 240 941 or (e-mail) info@wiley-capstone.co.uk

Contents

To Don and Julia.

Introduction to ExpressExec

ExpressExec is 3 million words of the latest management thinking compiled into 10 modules. Each module contains 10 individual titles forming a comprehensive resource of current business practice written by leading practitioners in their field. From brand management to balanced scorecard, ExpressExec enables you to grasp the key concepts behind each subject and implement the theory immediately. Each of the 100 titles is available in print and electronic formats.

Through the ExpressExec.com Website you will discover that you can access the complete resource in a number of ways:

» printed books or e-books;
» e-content – PDF or XML (for licensed syndication) adding value to an intranet or Internet site;
» a corporate e-learning/knowledge management solution providing a cost-effective platform for developing skills and sharing knowledge within an organization;
» bespoke delivery – tailored solutions to solve your need.

Why not visit www.expressexec.com and register for free key management briefings, a monthly newsletter and interactive skills checklists. Share your ideas about ExpressExec and your thoughts about business today.

Please contact elound@wiley-capstone.co.uk for more information.

Introduction

What is the role of managing technology in the modern world of business and institutions? This chapter considers the changing nature of technology and management practices. It includes:

» an argument outlining why is it crucial to become a tech-savvy manager who uses technology strategically, rather than allowing technology to run you and your organization; and
» advice on how to deal with the psychological issues that technology raises.

''If they don't do the right things, technology by itself doesn't help much. You might have wonderful technology, but has anything else changed? No. That is why some companies are in deep trouble.''[1]
Prabhudev Konana, Assistant Professor, The Center for Research in Electronic Commerce, The University of Texas

The Nasdaq may rise and fall. Dot-coms will come and go. But there is no doubt that anyone in business today will be managing technology of many sorts. It doesn't matter whether you work in fast foods, museums, hospitals, or industry – technology is as instrumental to your workplace as the product you manufacture or the service you offer.

Technology permeates all of your operations, all of your supply chains, in myriad ways. You rely on technology to communicate to co-workers, customers, suppliers, and shippers. You rely on technology if you manufacture a product to handle a wide variety of functions. You rely on technology to manage warehouse inventory and track shipments. And more and more, as we progress through the first decade of the new millennium, you will discover that many of your business functions will be networked and then serviced through that vast cyberspace storage bin called the World Wide Web.

There is no turning back. We operate in a techno world.

Managing an operation in a networked, Web-enabled environment requires new knowledge and expertise. You may find yourself forced to make technology purchases whether you feel equipped to do so or not. You must also supervise employees who are faced with technological malfunctions and incompatibilities. That means you deal with employee anger and frustration.

And you have to be able to field all of the technology-related issues while doing your primary job. If that's not hard enough, as managers you have to deal with your own personal frustration and the feelings of inadequacy and instability that new technology can produce in all but the most ardent techies.

Think of *Managing Technology* as your short guide through the minefield of operating a high-tech organization. The first aim of this book is to offer you ways that you can take charge of technology and move into the driver's seat, rather than have technology run you. To accomplish this task, you must become tech-literate, which means having a working knowledge of the technology pertinent to your operation. And you may have to change work habits to keep abreast of relevant technology developments. Knowing what sort of training you need and at what level will also be addressed, along with some tips for tracking technology change.

Secondly, this book is filled with advice on how to improve your management skills by recognizing and dealing with psychological issues that affect both you and your employees. Fear of change and computer phobia may sound silly, but they can actually demobilize the best technology efforts.

Managing Technology recognizes that managers are only human. Yes, there are days you'd love to pull the plug on your PC, toss your laptop out the window, and dump that cell phone that never seems to pick up calls when you most need them. Hopefully, the tips and advice that will permeate this guide will not only keep you from such extreme measures, but allow you to use the technology in your workplace to its best, strategic advantage.

NOTE

1 "E-Procurement: What To Do First," *Supply Strategy* online edition, June 2001.

Defining What it Means to Manage Technology

Most people know that technology matters to their organization, but they know little about how to manage it. This chapter outlines in more detail all the key elements that accompany managing technology. It includes:

» advice on training yourself and employees to recognize and comprehend the technology in your work environment;
» key technology trends in wireless communication, networking, and transportation/logistics, among others, that will affect your organization; and
» how to organize for technology implants, manage technology change, and select the technology that's strategically best for your operation.

"Dear PC/Mac God:
Give me the strength to accept there is little in Computer Land I can comprehend;
The courage to upgrade my equipment in a wise and timely fashion,
And the wisdom to call computer support personnel before I lose my mind."

The PC Surrender Prayer[1]

Tech-savvy managers like you recognize that you can't live without technology, but you don't always know how to live with it. Equipment isn't compatible. Expensive software solutions don't work or simply don't live up to their promise. Employees take out their anger and frustration on you when the system crashes and they lose that crucial report. How to manage in the world where plug 'n' play is still a dream? How to drive your technology decisions and not let the technology drive you? How to utilize technology so that it benefits your organization?

You don't need a degree in advanced computing or networking to learn to manage your organization in a networked, Web-enabled world. What many of you need is a change in mindset and alteration of daily work habits. Tech-savvy managers track technology change as avidly as they study the stock market. They build technology awareness and learning into their regular work schedule. They work on skills that will help them guide angry and frustrated employees and study the latest technology training techniques. And they recognize that this is a life-long. process.[2]

Getting yourself prepped to function in a high-tech world is the first step. Learning techniques for helping employees work with technology is the second. The following are key subjects and skills – from tracking technology trends to managing change and training – you need to manage a company or institution that relies on high technology.

TRAIN YOURSELF, FIRST

You can't expect employees to adjust to Word or run a program on Linux if you think the former is a dictionary element and the latter is related to a "Peanuts" cartoon. Managers have to know the basics of word processing and have some passing understanding of the technology that already exists in their organization. Let your IT department and vendors teach you the basics and then make sure that you book into your week sufficient time to learn about technology developments that could affect your organization. Remember, you're not aiming to become a techie – just to be tech-literate, tech-aware, and tech-savvy.

HUMAN SKILLS STILL COUNT

Being able to manage vast amounts of data and information accurately, let alone relay that information throughout a global supply network or post it on a B2B site, requires an entirely different set of skills than implementing a software solution. You and your employees need to know how to read for comprehension, analyze, gather, sort, and cull information for rapid and accurate dissemination. Otherwise, none of those technology expenditures will pay off.[3]

KNOWING THE TRENDS

You need to be able to track technology trends with an eye to what is here to stay. Without this knowledge you can't make cost-effective technology decisions, organize your company or institution to roll with technology change, or train yourself and your employees in the skills they'll need to survive in the New Economy. And you won't be able to apply technology strategically to meet business goals. You all face pressure from partners, customers, suppliers, and even government agencies, to move into the networked, B2B arena. The following are three major technology areas to tag while you make the transition over the next five to ten years:

» *Communication*: The communication world is practically seething with new wireless communication developments that aim to free you from offices, and even continents, while promoting customer relationships rather than merely improving customer services. To make this happen, suppliers in the world of technology and standards are working on convergence of telecommunications and the Internet. Sometime in the next five years you will be accessing voice, video, and data over your cell phone.

» *E-commerce and B2B*: Take your networked operation, plug it into the Web and you have moved from a supply chain to a supply network. From this base you will be able to access Web services in the most cost-effective and efficient manner possible and practise B2B. To get to this stage it's important to understand the principles of moving information electronically within an organization – and those technologies available today to do so – before looking at the more futuristic approaches to developing a supply network. You need to know about networking, integration, and the ways they interact with new Web technologies.

» *Transportation and logistics*: Because transportation and logistics are so crucial to maintaining material goods and product flow in this New Economy, these industries are using technology in enormously creative ways. Basic computing, satellite, radio frequency, bar codes, cellular, X-ray, and now the Internet – just about every technology imaginable – is being combined to literally and figuratively drive the supply network. It would not be possible to source goods, track shipments, inventory goods, or send freight bills on a real-time basis – let alone on a global basis – without advanced technology.[4]

TRACKING CHANGE

Tech-savvy managers want to be able to track technology change like the experts. To beat the analysts you need to know about technology user groups, conferences, and the journals and publications the experts cite as the best source of technology news and information. And the really tech-savvy managers know how to play in the world of standards, where technology is developed and harmonized for global use. They know about the organization of both the international standards arena and the world of high-tech consortia, as well as the key working groups and committees developing and standardizing the technology that affects their operation.[5]

MANAGING CHANGE

Technology will make change. That's the most important factor strategic decision-makers, professionals, and entrepreneurs have to know and accept to stay competitive while managing networked/B2B-oriented organizations. Change is confusing, difficult, and often painful. But, as all of you know, there's no way to avoid technology change. Accepting the change technology brings, rather than resisting it, will help ease you into a world in which technology serves you and your organization, rather than you serving it. And knowing the emotional stages that both you and your employees will experience will make you better equipped to promote change within your organization and reap technology's benefits.[6]

ORGANIZING FOR A NETWORKED, B2B FUTURE

As tech-savvy and seasoned managers know, organization and set-up are half the battle. If you take the time up-front to establish human procedures and rules, processes and communication channels, you will find the job of managing in a networked, Web-enabled world far easier and you will more than earn back the cost of technology implants. Knowing how to organize your operation to function in both three-dimensional and virtual worlds, promote communication and accuracy, and operate electronic equipment and computing is key to success.

As always, business strategy should be top on the list when dictating a technology course. Management issues are also key. You will want to explore how information flows in your company, and the sorts of skills your employees have – or lack – to handle the tasks of working in an networked, Web-based fashion. Ensuring that employee schedules are flexible to meet the demands of this new world is also crucial.[7]

MAKING TECHNOLOGY CHOICES

Even if you have no intention of becoming a techie, you may find yourself forced into making technology selections that have huge ramifications for your company or institution. You don't have to be a skilled computer programmer to learn how to make wise technology choices, and to work with your IT (Information Technology) department on this process or with technology vendors. The key to a wise, cost-effective approach to this process is knowing and sticking to business priorities while keeping an eye out for technology that will promote goals and be both flexible enough to grow with you and scalable enough to be compatible with future technology developments.

Learning how to conduct IT assessments is part and parcel of this process. Basically, an IT assessment is a method for both cataloguing technology and comparing what you have in-house with the benefits or deficits it produces. This information is used as a benchmark device against future technology purchase. Managers starting from scratch need to spend some quality time researching what's on the market. As noted earlier in this chapter, that may mean a learning curve and continued information gathering.

KEY LEARNING POINTS

» You don't need a degree in advanced computing to manage an organization using advanced technology, but you do need to train yourself in the rudiments of the technology you use.

» Training for yourself and employees is an ongoing process that may require changes in work habits.

» Being able to read, write, and analyze are key to using the data and information that technology provides in a strategic fashion.

» Knowing what trends are here to stay in the areas of Internet/ Web, wireless communications, e-commerce, business processing, and transportation and logistics will make you a stronger manager.

» Boning up on how analysts, consultants, and vendors operate, plus coming into meetings with them with a basic knowledge of the technology they're promoting or assessing, is crucial to taking charge of the technology you use.

» Organize your operation for maximum information flow first, and then implant technology later.

» Prepare yourself and your organization to function in a networked, Web-enabled environment.

» Conduct an IT assessment of the technology in your organization before making new purchases.

NOTES

1 Zuckerman, A. (2001) *Tech Trending*. Capstone, Oxford, p.16. Adapted from *When Bad Computer Things Happen To Good People*, a work in progress by Amy Zuckerman and Izzy Gesell.
2 *Ibid.*, p.3.
3 *Ibid.*, p.14.
4 *Ibid.*, pp.9–10.
5 *Ibid.*
6 *Ibid.*, p.11.
7 *Ibid.*, pp.11–12.

The Evolving Networked, Web-Enabled World

Technology is as old as the abacus, but much of the technology used in the workplace today dates from the post-war era. This chapter examines in more detail the technology that has helped produce a networked, Web-enabled world. It includes:

» how we got to networks, cell phones, and the Web;
» the transition from mainframes to servers, from a supply chain to a supply network; and
» from fiber optics to microphotonics.

"Now the issues are when will companies begin living B2B and how to define what 'living B2B' means. Is it 50% of the transactions? Some experts believe this world is 10 years off, some 20 and even 30 years. No one knows the answer to that question."[1]

Kevin Fitzgerald, editor-in-chief and associate publisher of
Supply Strategy

Any of you who started your careers in 1970 have witnessed unimaginable technology change in the course of the last 30-plus years. Outside of major corporations and major urban centers, technology in the workplace hadn't evolved much since the mass adoption of the telephone 100 years ago. In the workplace of 1970, the primary piece of communication technology equipment was the telephone. Urgent information was relayed by telegraph or courier. Typewriters had advanced from manual to electric. Computing was restricted to mainframes whose main purpose was massive number crunching.

Contrast that world with today's technology. Phones are still staple work items, but out of the office, more and more of you have switched to wireless cell phones or what some call "mobiles." Most companies, even the smallest, have an Internet connection and are contemplating developing Websites to advertise their goods and services in cyberspace. Less advanced operations rely on fax machines, which came into mass use in the 1980s.

Thanks to the invention of the computer chip and semiconductors, mainframe computers have largely – if not entirely – given way to PCs. Servers allow you to network an organization and allow participants to access databases, as well as the Web. It's possible to source and procure goods over the Web and download software onto Web servers. And shipments can be tracked in real-time via the Web and transportation/logistics information relayed quickly between carrier, customer, and shipper.

Fast forward five to ten years. Experts agree that we are evolving into a networked, B2B (business-to-business electronic commerce) world where information will flow instantaneously throughout an organization, to outside training partners, vendors, and collaborative trading groups. Some visionaries are talking about peer-to-peer (P2P), where millions of computers will be networked on the Web for data-mining and information sharing without the need for Web browsers or filters. Information will be communicated anywhere in the world, and using Web-enabled cell phones you will be able to access whatever data or information you need to operate a company or institution while on the fly.

HOW WE GOT TO NETWORKS, CELL PHONES, AND THE WEB

Can you remember your first e-mail account password? The first time you bought a PC or laptop? The first time you signed up for a cell phone service? Probably not, even though PCs have only been available on the mass market for about 15 years or so. The world of electronics, information technology, and telecommunications has evolved so rapidly it all seems like a blur to those of us who started our careers relying on phones and maybe a fax machine. But it's taken the combined forces of many developments – in everything from computer chips and semiconductors to telecom switches and cell wireless technology – to breed the networked, Web-enabled world many of you take for granted today.

From mainframes to servers

No one would be talking about networking and integrating an operation without the invention of the operating systems and PCs. And no one would be talking about PCs without the development of the computer chip and semiconductor. All of this technology has evolved since the late 1960s, with the introduction of the UNIX operating system that could network minicomputers and large microcomputers. Thanks to major advancements in the chip world, computing power has grown exponentially, just as the size of the actual box has shrunk to a desk top. The average server, which can store as much data as the mainframe of 40 years ago, is about the size of a pizza box.

The computing world we know today, which is often based on the client/server setup, has been prevalent since the 1980s and has relied on the development of stronger and stronger chips and software applications that can handle a wide variety of business functions.

By the mid-1980s, the cost of computing fell to prices affordable to even average consumers. By the 1990s, many large companies and institutions had evolved ways to network and integrate their operations through database software applications downloaded into servers rather than mainframes. (Mainframes are still useful for large-scale computing.)

This move to networked organizations is still very much a hot trend today. Only the largest companies and institutions, and the most cutting-edge smaller organizations, are fully networked, meaning that core business functions are stored and available throughout your operation to all employees with access. Many of you operate with a hodge-podge of technology that will require either replacement or software integration tools to provide a networked environment. Networking is key to any smart manager keen on staying competitive in a Web-enabled, B2B world. Even if you choose to outsource many computing functions, you still need your operation to run flexibly and efficiently and to create databases to store information.

HOW TECHNOLOGIES ARE EVOLVING THE SUPPLY CHAIN INTO A SUPPLY NETWORK

Internal organizational communication

» Supply Chain Technology – Telephones and faxes.
» Supply Network Technology – E-mail, company/institution intranets, and Web-enabled cell phones allow tech-savvy managers to share information and communicate while on the go.

Business process communication

» Supply Chain Technology – Business data and information flow from department to department via e-mail and enterprise resource planning (ERP) software.
» Supply Network Technology – New standards for Business-to-Business (B2B) communication will allow the transmission of everything from invoices to reports throughout a supply network. Peer-to-Peer (P2P) software will allow you to network with millions of computers worldwide without the need for a browser.

External communication

» Supply Chain Technology – Telephones, faxes, and electronic data interchange (EDI) are being aided by the Internet as a transmission tool.

» Supply Network Technology – New standards will eventually eliminate EDI messaging sets and allow company and institutional computers to communicate directly.

Computing

» Supply Chain Technology – Mainframe computers dominate to store massive amounts of data.

» Supply Network Technology – Networked systems that rely on servers are supplanting mainframe computers. Servers are being connected to Web browsers to allow data and information to flow directly to cyberspace. Peer-to-Peer software (P2P) may make the browser unnecessary, if not obsolete.

Purchasing

» Supply Chain Technology – Phone, faxes, and EDI.

» Supply Network Technology – The Web is becoming the place where purchasing managers can source goods electronically. As Web languages evolve, computers will handle purchasing directly.

Inventory management

» Supply Chain Technology – Computerization, aided by EDI and now the Internet, has allowed information to flow from warehouses to manufacturers and shippers so that all parties have accurate, real-time inventory counts and can manufacture and ship just-in-time.

» Supply Network Technology – Advancements in radio frequency and bar code technologies, and the introduction of remote messaging technologies, allow for instantaneous stock counts with the wave of an RF wand.

Shipping

» Supply Chain Technology – Phones, faxes, and EDI are coupled with computing, as well as satellite, global positioning, and

cellular technologies, to allow for near real-time global tracking and tracing of goods.
» Supply Network Technology – The introduction of the Internet as a data transmission tool means that real-time information is now available. Intelligent transportation systems (ITS) coupled with computers are moving goods more quickly along highways, promoting safer transportation and providing pass-through capabilities at border crossings. And transportation and logistics services – including tracking real-time tracking information and software products – are now available on the Web.[2]

From fiber optics to microphotonics

There's a hidden world of telecommunications switching that may very well revolutionize the way you all do business. Just as fiber optics has allowed for greater information transmission via myriad glass cables replacing traditional phone lines – paving the way for Internet and Web use – microphotonics technology is expected to hugely increase the amount of data that can be transmitted via telecommunications networks. And microphotonics is expected to resolve the problem of the "electronic bottleneck" that is predicted as multimedia hits the Net in a big way.

Microphotonics is based on an "optical switch," which directly routes information in the form of photons of light without the need to convert to electronic signals. Optical switches are infinitely smaller than anything now on the market. And they can outperform existing devices, orchestrating 32 beams of light in less than one-hundredth of a second. Companies from Nortel Networks to Lucent Technologies have reportedly been pouring millions into optical switch research and development, even going as far as to buy startup companies working in this arena.

Besides their speed and capacity, experts consider optical switches an advancement that promises to reshape global communications networks. Some researchers believe optical switches foretell the development of a tiny but complex optical device that may change how data is transmitted in everything from communication systems to computers. The first application of photonic switches will be to eliminate the costly electronic switching that makes up the fiber optic backbones of today's communication networks that run from city to city. Increased Internet use is stretching today's fiber optic infrastructures.

Future generations of photonic switches are expected to become smaller, offer more capacity, be cheaper, and be "smart" so that they have the capability of reading the messages carried in them. This is particularly important to Internet design, where being able to read the messages attached to switching nodes will allow for better quality service, quicker relay, and maintenance ease. The new optical switches will actually replace Internet routers. Experts also see the use of photons as the key to faster computer chips, resulting in smaller transistors on silicon wafers and greater computing power.[3]

From fixed phones to wireless communication

It's almost impossible to remember the days when phones were attached to plugs in the walls and conversations were constricted to the length of the cord you could buy at a phone appliance store. The so-called cordless phones evolved first and were hot consumer items as early as the 1980s, or even earlier in some environments. By the 1990s, cell phones started to arrive on the scene. Business people were heavy users in the beginning, but nowadays cell phones are so ubiquitous that you see them everywhere from airports to obscure villages in Tuscany.

Standards were published in 2001 that make it possible for manufacturers worldwide to produce mobile phones that are operable anywhere on earth. These are the so-called standards for the Third Generation (3G) Mobile Communications that are developed by two groups – the Third Generation Partnership Project (3GPP) and 3GPP2. The new generation of specs support global roaming and access to the Internet and other multimedia services at speeds up to two megabits per second (Mbits/sec). This is in addition to the voice services already familiar to mobile users.

Applied to mobile phone products, standards developers report that the new family of specs offer better voice quality, better security systems built in, and a much higher bit rate than in 'second generation' technology. How you use the data is up to the imagination of the application and service provider, because there is more than one way to accomplish better voice quality, better data capabilities, and all that the futuristic third generation services offer.[4]

Already, more advanced versions of 3G standards are in the pipeline. One emerging trend in communications involves convergence of telecom and Internet capabilities. (See Chapter 6: The State of the Art) Working out the kinks in Internet/telecoms integration will hopefully improve those pioneering Web-enabled cell phones that are now on the market. Users find service erratic, handsets clumsy, and Web searches via wireless to be slow.

The emergence of the Internet/Web

First developed by the military and adopted by academics, the Internet is built around packets – strings of data that find their destination by hopping between local and regional "nodes" where the mesh of transmission lines that form the Internet intersect. There are thousands of packets that travel separately from node to node, each carrying with it an Internet protocol (IP). This is the infamous IP that telecoms engineers are always speaking about, and that allows an address to be read by electronic switches called "routers" at each node. The IP address tells the routers where the packet is going, and then the routers – which must analyze each packet – forward them on to the appropriate node.

This process is repeated until the individual packets that make up a message have arrived at their destination and are rearranged. Experts say it's this forwarding arrangement that makes the Internet resilient. If a node goes down or is temporarily swamped, neighboring routers simply divert the flow of packets around it. There is a concern that the size of the routing table – the table of Internet destinations – is growing too fast and is changing too frequently. There is much discussion taking place right now about how to address this issue, but there are no immediate solutions in sight. This is something that should concern you and an issue you should certainly tag. Always remember that the Internet is now in integral part of your business; one you can't afford to lose.[5]

Mass use of the Internet didn't materialize until the mid-1990s with the evolution of an Internet service provider industry. Within a year or so the World Wide Web had developed to the point where millions of people were posting information in what amounts to a major cyberspace storage bin. Nowadays, the trend is to make services

"Web-centric," meaning that they can be accessed through Websites. From books to logistics services, it's possible to purchase many goods and services in cyberspace today.

But the more of you that depend on the Internet/Web to run your operations, the more that performance, capacity, and bandwidth become all important, not to mention reliability when it comes to offering fool proof customer service. So far, horror stories about bandwidth crunches haven't come true, and customer fears about security haven't materialized, but those are fears that tech trending managers say are quite real.[6]

As the numbers of Web-based operations escalate, these sorts of concerns will grow as well. The good news is that a great deal of work is taking place in standards organizations to improve Internet capacity and assure quality service as usage spirals upwards. Equally good news for tech trending managers eager to see B2B become a daily reality are developments in Web markup language that will provide the capabilities required to place catalogues on the Web and hook buyers to sellers on the global scale.

For example, Internet maintenance committees at the Internet Engineering Task Force (IETF) are working hard to assure quality Internet service. IETF is a volunteer-run consortium and a subset of the Internet Society (ISOC), based in Reston, Virginia. Keeping the Internet working in the midst of rapid growth is a major concern of standards developers at IETF, who are working on everything from telephony to Internet security and development of e-mail standards.

TIME LINE

The following are key developments in the post-war history of technology that tech-savvy managers should know, or at least be able to drop at pertinent times:

» **1946**: The ENIAC, the world's first practical all-electronic digital computer, is developed.
» **1948**: Scientists at Bell Telephone labs, including William Shockley, Walter Brattain, and William Bardeen, invent the transistor (solid state amplifier).

» **1957–1958**: Jack Kilby, one of eight electronics engineers and physicists from Fairchild Semiconductor, invents the first integrated circuit.

» **1960s**: The US Department of Defense develops ARPANET, which becomes the Internet. It was originally intended to be a network of government, university, research, and scientific computers designed to enable researchers to share information in the event of a nuclear attack.

» **1969**: Developed by AT&T engineers Ken Thompson and Dennis Ritchie, the UNIX operating system is introduced. It can handle multitasking and was ideally suited for networking minicomputers and large microcomputers.

» **1971**: A team from Intel, under the leadership of Federico Faggin, develops the first microprocessing chip.

» **1972**: Philips and MCA develop the optical laser disc.

» **1974**: Motorola produces its first microprocessor chip, the Motorola 6800, used in the original Macintosh computer.

» **1975**: Altair Microcomputer Kit becomes the first personal computer available to the general public.

» **1977**:
   » Radio Shack introduces the first personal computer with keyboard and CRT display.
   » Apple Computer begins delivery of the Apple II computer.

» **1984**: Apple Macintosh computer is introduced.

» **Mid-1980s**: Artificial intelligence (AI) develops as a separate discipline from information science.

» **1987**: Bill Atkinson of Apple Computer develops the HyperCard, which makes hypertext document linking possible.

» **1991**:
   » CD-ROM storage technology and computer-based search engines are now available to the public.
   » IBM, Motorola, and Apple introduce the PowerPC Chip.
   » Linus Torvalds introduces the open-source operating system called Linux.[7]

» **1992**: Microsoft introduces Windows 3.1, bringing Apple-type easy graphics and computing to millions of PC users.

» **1993**:
  » Fifty World Wide Web (WWW) servers exist as of January. President Bill Clinton puts the White House online with WWW addresses for the president, vice president, and first lady.
  » Mosaic, the first graphical Web browser, is released by the National Center for Supercomputing Applications (NCSA) at the University of Illinois.
» **1994**:
  » Marc Andreesen and James H. Clark found Netscape Communications and release Netscape Navigator browser software, which provides an easy, point-and-click method of navigating the Internet.
  » Microsoft, Hewlett-Packard, U.S. West, Telstra, Deutsche Telekom, NTT, Olivetti, Anderson, and Alcatel join forces in an effort to develop the hardware and software necessary for the actualization of interactive television.
» **1995**:
  » Apple finally allows other companies to clone the Macintosh computer. This proves to be a little bit too late for it to become the market leader.
  » A number of Internet-related companies go public. Netscape has the most successful initial public offering (IPO), opening at $28 per share and closing at $58 per share.
  » Intuit, the maker of the financial software Quicken, announces that it is working with 19 financial institutions, including American Express, Chase Manhattan Bank, and Wells Fargo, to develop on online link that will let customers with modems dial into their accounts.
  » "Java" and "telephony" are the buzzwords on the Internet. Java allows small applications, called applets, to be run on Websites, expanding the capabilities of the World Wide Web. Telephony lets users talk to each other over the Internet without paying long-distance telephone charges.[8]
» **1996**: Linux 2.0 is released, which makes it accessible to many users.
» **1998**: Microsoft releases Windows 98. Some US attorneys try to block its release, since the new OS interlaces with other programs such as Microsoft Internet Explorer and so effectively closes the market of such software to other companies.

» **1999**:
  » Linux kernel 2.2.0 is released. The number of people running Linux is estimated at over 10 million, making it not only an important operating system in the Unix world, but an increasingly important one in the PC world.
  » AMD releases K6-III 400MHz version, with samples of the 450MHz chip going to original equipment manufacturers (OEMs). In some tests it outperforms the soon-to-be released Intel PIII.
  » Apple releases the PowerMac G4, which it claims to be the first personal computer to be capable of over one billion floating-point operations per second.[9]

KEY LEARNING POINTS

» We are rapidly moving into a networked, Web-enabled world where communication is wireless and many business and institutional functions will take place in cyberspace.
» The invention of the integrated circuit that led to the development of superchips is the basis of the computing world we know today.
» In the last 30 years, the world has largely moved from mainframe computing to utilizing servers that operate on database applications, all of which is making it possible to network organizations.
» The evolution of fiber optics and more advanced telecommunication switches has vastly promoted use of advanced technologies.
» Wireless communication is evolving so that it will be able to handle Internet/Web traffic, freeing managers and employees alike from offices.
» As the Internet/Web evolve, more and more business functions will be handled in cyberspace rather than in the three-dimensional world.

NOTES

1 Zuckerman, A. (2001) *Tech Trending*. Capstone, Oxford, p.68.
2 *Ibid.*, p.69.

3 Ibid., p.33. This material is adapted from Fairley, P. (2000) "The Microphotonics Revolution," *Technology Review: MIT's Magazine of Innovation*, June/July. Additional source is Gerald Peterson, global telecommunications standards director, Lucent Technologies.

4 Zuckerman, A. (2001) *Tech Trending*. Capstone, Oxford, p.35. Sources are Asok Chatterjee, chairman of T1's Technical Subcommittee T1P1, the Alliance for Telecommunications Industry Solutions (ATIS), Washington, DC, and Gerald Peterson, global telecommunications standards director, Lucent Technologies.

5 Ibid., p.47. This material is adapted from Fairley, P. (2000) "The Microphotonics Revolution," *Technology Review: MIT's Magazine of Innovation*. (online edition.)

6 Zuckerman, A. (2001) *Tech Trending*. Capstone, Oxford, p.48. Source is Peter Schwartz, Descartes Systems chairman and CEO.

7 Source for 1946–1991 is "Major developments in Computing since 1946," at www.AskJeeves.com

8 Source for 1991–1995 is Computer Tech On-Line Copyright © 1995–2001.

9 Source is "My Homepage" by Stephen White, 1996–2001, at www.ox.compsoc.net/~swhite/timeline.html

The E-Dimension

Technology, particularly electronic commerce, poses new challenges for tech-savvy managers. This chapter explores these key issues. Among them:

» how to prepare for a B2B environment;
» the technology you need to practice rudimentary e-commerce; and
» what e-commerce model is right for your operation.

"I wanted to use technology to get a piece out of what's a $32 bn industry. The Internet seemed the way to go."

Steve Odzer, chairman, chief executive, Global Supply Net[1]

No tech-savvy manager operates a company or institution today without considering whether to conduct business on the Web, what's commonly called electronic or e-commerce. A revolutionary concept only a few years ago, e-commerce is starting to permeate the global business economy. Consider the following examples.

A cultural tourism organization in Cape May, New Jersey uses the Web to promote its services and boost ticket sales. A wooden-ship training center in Victoria, British Columbia advertises e-mail cards as a come-on to support its services. A pork rind manufacturer in Georgia is developing a Web presence to improve sales. And larger companies, from the Big Three auto makers to aerospace manufacturers, are all developing cyberspace exchanges to cut down on purchasing and operating costs.

You can be certain that whatever trend really delivers and promotes cost-cutting and profit-making is here for the long term. And e-commerce holds the promise of cost savings, increased productivity, and more. Purchasing experts are predicting that fully-evolved business-to-business e-commerce (B2B) systems will provide productivity improvement, reallocation of procurement resources, material cost reductions, decreases in the need for invoicing, electronic expense report savings, and cycle time reduction savings. In the B2B model of the future, computers conduct business on your behalf.[2,3]

But as tech-savvy managers know, B2B is still in its infancy. Most B2B is taking "indirect production supplies" rather than in the purchasing of "direct production supplies." Indirect supplies are those needed to maintain and operate a facility that don't necessarily relate to direct production of a product. What is being purchased today involves items such as lighting or electrical and plumbing supplies that don't affect a supply chain network and don't involve much risk. These are high-volume, low-cost items that every company needs to purchase and where online purchasing can eliminate waste.

A lot of B2B activity also centers on gathering information for sourcing to obtain lower prices. For companies to move onto the direct purchase side means eliminating the element of risk that offers. Buyers and suppliers will have to gain confidence that B2B produces results, and there will have to be management buy-in. As B2B emerges from the fantasy stage to reality, a number of models are evolving. Of these, the Application Service Provider (ASP) format appears to be really taking off. The Phillips Group predicts 56% growth in this model among large companies by 2004.[4]

Infancy or no, almost every industry – from metals to tourism – is evolving an e-dimension. And elementary e-commerce practices are spreading like kudzu throughout many common business functions. There's e-purchasing, e-sourcing, e-logistics, and e-transportation, where tracking shipments via the Web is pretty much commonplace. As tech-savvy managers, you can't afford to either ignore e-commerce or avoid preparing your organization for a networked, Web-enabled world.

MOVING INTO AN E-COMM MODE

So you have fully automated and integrated whatever business functions that make sense for your overall strategy, and some of you have a fully networked operation. You now face that looming e-comm world. Before you buy any new equipment, software applications, or even design a Website, it's imperative to decide if e-comm is right for your organization at this juncture, or if it's five to ten years away.

Remember, it's one thing to purchase a book at Amazon.com and quite another to practice to move your operation into cyberspace. Consider e-procurement for starters. On the buy side, you will need Web access and strong search engines, at the very least. On the sell side, tech-savvy managers will need Web presence, e-catalogs and electronic order fulfillment. Most companies and institutions have long ago computerized the buy side and sell side functions, or at least significant parts of these processes. For those sort of tech-savvy managers the challenge is to further speed up and streamline the entire procurement cycle by more closely linking buyers and sellers.

At its best, e-procurement provides organizations with the ability to automate routine practices and tighten supplier involvement and integration. Tech-savvy purchasing organizations are moving beyond this stage and working on developing buy/sell collaboration between suppliers and vendors, creative risk sharing, online auctions, complexity reduction, and management of complex channel relationships. For example, General Electric has created a "Trading Process Network," an electronic system that sends out requests for quotations – along with required drawings – digitally to global vendors. GE managers cite 50% cycle time reductions and lower material acquisition costs among the benefits. Cisco Systems claims its e-procurement processes have reduced acquisition costs by $105 per transaction.

What's good for GE and Cisco may not be what's good for smaller organizations. Experts say e-procurement only works when buyers and suppliers have the right management systems and technology in place. Your operation has to be able to support a switch to electronic buying and selling, suppliers must be able to handle new information processing requirements, and all systems must be compatible. And you better be able to manage content. As will be noted in sections that follow, managing data in a Web environment is far more complicated and detailed than working in print.[5]

WHAT A DOT-COM STARTUP TAKES

A fledgling Philadelphia-based dot-com named BioSupplies.com has moved from creating a transaction-based e-commerce product for life science researchers in academia to creating software – under the new name Tangerine Technologies – that's extensible and accessible to other verticals. Founders estimated they needed gross sales of $200 to $400 mn annually to make a go of it and the pathway wasn't there on the $4.5mn they raised in venture capital.

Some of the original BioSupplies.com team has cooked up a new dot-com – Varro Technologies – that involves sharing information on the genome with academics and research institutes. This would involve sale of a software product that allows people to access new content. Subscription fees would be charged. To move this concept out of startup would require an estimated $40 mn. Founders say it's possible to create a prototype dot-com that's scalable for a few million dollars. But the minute you put the model on servers, which requires specialized hardware to support data loads and have the appropriate software, then you're kicking the costs into the millions.[6,7,8]

KNOWING WHAT E-COMM TECHNOLOGY YOU NEED

Think of your basic networked operation with its servers hosting a variety of databases and then move the whole shebang onto the Web. Here's the sort of technology tech-savvy operations and managers will need to make the leap to a Web-based, B2B world.

On the hardware side you're going to need to connect your company or institution's network to a Web server that, in turn, connects to the Internet via a Web browser. Web servers are servers that contain specialized software geared to the sending and receiving of data; the computer version of a switch. On the software side you will need a database to drive the system, and that may require application and development tools, authoring and maintenance tools, and network and integration tools. Now that you're out on the Web, data security is a main concern and you'll have to decide whether you want to encrypt your data, post firewalls, or work in Secure Electronic Transactions (SET).

You will want to consider implanting a content management system that will manage data presentation, enable distributed updating, and quickly post customized changes. Knowledge management tools will allow you to post something as simple as Frequently Asked Questions (FAQ) on your website, or something as complicated as the ability to provide information to a client in need of online customer support. The more you edge out onto the Web, the more solutions you will both want and need. There are customer relationship systems, systems for ad management, and portal development software. If you're involved in e-tail you may want a taxation software package, and if you go global you'll want software to manage everything from duties and tariffs to exchange rates.[9]

What e-comm model is right for you

Then there's the question of what you actually want to do out on the Web. It may be as simple as sourcing goods or as complicated as conducting all your business transactions in cyberspace. You will have to decide what sort of Web-based marketplace or exchange to enter. Maybe these concerns are several years out, but there's no doubt that the world is lining up to produce a Web-based environment. Here are some e-comm models available today.

» *Online marketplaces:* Companies like Ariba, SAP, Oracle, and Ford have announced online marketplaces that offer a collection of catalogs that allow buyers to source and purchase. The more sophisticated sites have provisions for customer-specific pricing or total site volume pricing. They also offer content to attract users, such as vertical industry news, articles, and even entertainment. And they support some degree of integration so that orders and related documents generated on the site can be automatically downloaded into a users' financial systems. That's one reason that ERP vendors like SAP and Oracle are setting up their own marketplaces.

» *Trading communities:* In the e-commerce sense, companies use the term "trading community" to describe a group of buyers and sellers that work in the same market-space, sharing common needs, information, and business partners. For example, companies that buy and sell on the chemical supplies site, Chemdex, would be considered a trading community.

» *Buy-side procurement systems:* Still pretty much limited to Fortune 100 companies with lots of money to invest in end-to-end systems, this model allows a purchasing department to direct their end users to preferred suppliers. Companies like Commerce One, Ariba, and Intelisys are providing these sorts of applications. The major drawback is cost, which can range from $250,000 to $5 mn.

» *Service bureaus:* Also known as application service providers or infomediaries, service bureaus basically provide an aggregation of suppliers and distributors of indirect supplies. This allows buyers to access indirect supplies from one location. Service bureaus represent the most common form of B2B sites.

» *The sell-side model:* This is the least sophisticated form of B2B and usually involves a supplier setting up shop on the web with his own products or products available for purchase. This model replaces phone sales, but doesn't add much other value. An example is Staples putting up a Website and selling its products.

» *Exchanges:* These host many-to-many buyer-seller relationships for more commodity-type items. A common type of exchange involves transaction fees to the sellers. In other words, someone brings buyers and sellers to a Website, facilitates one-on-one purchasing and takes a piece of the transaction.

» *ASP models:* These are software-rental arrangements where the vendor hosts and maintains a piece of software and simply sells the right to access the software. Many see ASP models taking over from the days of buying a piece of software for local installation, when ongoing upgrades were limited. When ASP takes over, or so the conventional wisdom goes, it will be much cheaper and more scalable to rely on a service provider for upgrades and maintenance. ASP is also supposed to promote the growth of various types of intermediaries facilitating communications/transactions between buyers and sellers. Product data preparation, maintenance, and dissemination are emerging as basic needs in the marketplace. These solutions are often based on ASP models.

» *Sell-side auctions:* This is where multiple buyers compete for products and services with upward price pressure.

» *Buy-side auctions, or reverse auctions:*. Buyers of indirect materials and services turn to buy-side auctions or reverse auctions because sell-side doesn't work for them. In this model, you, the buyer, get to auction off your requirements. Sellers get the chance to "bid" on fulfilling your need, and at a pre-determined point in time, the auction is "over" and the cheapest supplier is given the chance to earn your business.[10]

BEST PRACTICE CASE STUDY – GLOBAL SUPPLY NET[11]

A distributor of janitorial supplies, Global Supply Net of New York City is an old-economy company with a new-economy approach – selling Web technology to other distributors.

Company officials started exploring e-commerce potentials as early as 1999. A $100,000 study from Harvard Computing Group Inc. in Westford, Massachusetts was commissioned to outline how the Company could sell janitorial supplies on the Web. Because of the fragmented, localized nature of the industry – shipping costs often outweigh what distributors can earn in margins on goods like toilet paper and bleach – owners decided to provide technology to other distributors.

First, they raised $8 mn to finance the development of an e-commerce arm. To become a so-called B2B facilitator has cost several million dollars. The Website – which cost $2 mn to set up – includes product information, easy ordering, industry news, and safety sheets to meet federal regulations. Anyone can enter the site through www.globalsupplynet.com or www.jancentral.com.

Besides the Website expenses, 20 employees were added to run the new e-business at a cost of $1.5 mn a year. Then there's the estimated $1 mn a year for technical support and maintenance. To sustain projected growth, Global Supply Net technicians have recently upgraded the company's computer servers. Exodus Communications of Santa Clara, California – which hosts major companies like Amazon.com – hosts Global Supply Net's servers for $6,000 a month.

Distributors of janitorial supplies that want an immediate Web presence can buy a license to the site for $3,000 a year and customize it to their own needs. The technology behind Global Supply Net reportedly allows other distributors to build catalogs and Web sites and do B2B for their own customers. So far, about 100 distributors have syndicated the site and are selling to 300 end users. Owners predict they will be profitable by 2002.

KEY LEARNING POINTS

» E-commerce, or buying and selling on the Web, is the wave of the future.

» This practice, particularly business-to-business e-commerce, is still in its infancy, but it's a process that tech-savvy managers must learn more about.

» To take full advantage of e-commerce potentials, you should have a fully networked, integrated operation, so that all back office (financials) and front office (sales and ordering) functions are handled electronically.

» E-procurement, or purchasing via the Internet/Web, is rapidly being adopted by many large companies and holds the promise of huge cost containment.

» Networked companies will require a Web server that connects to the Internet via a Web browser to practice e-commerce, a database application to drive the system, and security in the form of firewalls or encryption to protect sensitive data and information.

» Content management and customer relation systems are advancements you may want to consider once your basic e-commerce setup is working.

» Choosing the e-commerce model that's right for your organization means knowing a lot about the ways you want to market your products, as well as knowing the difference between a Web marketplace, auction, exchange, or service bureau. These various models will have a major effect on what sort of buying or selling you conduct.

NOTES

1 Zuckerman, A. (2001) "Making the Leap with Mops, Flowers and Pork Rinds." *New York Times E-Business Special Section*, February 28, 2001, p.3.

2 Zuckerman, A. (2001) *Tech Trending*. Capstone, Oxford, pp.248–9.

3 Mazel, J. (2000) "How to Start an E-Procurement Initiative: Three Experts' Advice," *Supplier Selection & Management Report*, July 1.

4 Zuckerman, A. (2001) *Tech Trending*. Capstone, Oxford, p.248.

5 *Ibid.*, pp.248–9.

6 *Ibid.*, p.249.

7 DeLano, C. & Tibbens, R. (1999) "Is an E-Procurement Solution Right for You?" *Mercer Management Consulting*, June 1, appeared on PurchasingCenter.com in the summer of 2000.

8 Source: Jay Venkatesan, co-founder of BioSupplies.com and CEO of Varro Technologies, Philadelphia, Pennsylvania.
9 Source: Cunningham, M. (2000) *Smart Things To Know About E-Commerce*. Capstone Books, Oxford, pp.53–71.
10 Zuckerman, A. (2001) *Tech Trending*. Capstone, Oxford, pp.101–3.
11 Material is adapted from Zuckerman, A. (2001) "Making the Leap with Mops, Flowers and Pork Rinds." *New York Times E-Business Special Section*, February 28, p.3.

The Global Dimension

Any organization that buys and sells on the Web may find them-selves operating in a global environment. This chapter outlines the opportunities and pitfalls of moving into the global dimension. It includes:

» how e-sourcing, e-transport/logistics, and e-procurement can aid your operation; and
» how wireless communication will help you operate from anywhere in the world.

"From a business stand point we want to move to acceptance of click-and accept agreements on a global basis."
Richard Lucash, partner, Lucash, Gesmer and Updegrove LLP[1]

No company or institution operates globally without advanced technology. The ability to transmit information electronically via electronic data interchange (EDI) has been available for a number of years. But the Internet and the Web are creating not only cheaper and easier ways to transmit data and information worldwide, but new ways to conduct a wide variety of business functions. Other technologies – including cellular, radio frequency (RF), satellite, and intelligent transportation systems (ITS) – are also playing major roles in promoting global trade. Consider the following examples.

» The Web presents an instant global dimension for any business. With a simple implant of meta-tags to your company or institution Website you are suddenly connected to search engines that, in turn, connect you to the world of global trade.
» The Internet provides instant connection to just about everywhere in the world and drastically reduces the cost of communicating with remote offices or alliance partners. And the Net makes for cheap, easy transmission of huge amounts of data on a global basis.
» Combine the Internet and the Web, and tech-savvy managers can move into cyberspace to handle a slew of business functions from sourcing to logistics and transportation.
» New wireless capabilities, including the advent of Web-enabled phones, are allowing you to manage a company while on the fly and in foreign lands.
» Automated import/export systems are helping global traders push goods in and out of foreign countries far more efficiently and effectively.
» Want to move your goods across borders? Intelligent transportation systems (ITS) technology is already proving valuable to provide pass-through capabilities at border crossings.

Let's take a deeper look at the panoply of technology options available to tech-savvy managers itching to go global.

E-TRANSPORT/LOGISTICS[2]

The Internet and Web are having an enormous impact on the transportation industry. For example, the rush to cyberspace shopping has created a bonanza for expedited overnight carriers like FedEx, UPS, and others. These same carriers have been at the forefront of moving many of their business functions to the Web. And there's been a huge push to develop Web-based transportation dot-coms where all business transactions take place on the Web. However, only a handful of the pioneering transportation and logistics dot-coms are expected to survive into maturity.

Transportation visionaries see the Web being used more and more to better manage real-time operations. For truck fleets, the sort of optimization software that is already allowing the big companies to manage truck capacity better and ensure that all space is utilized each delivery will be available via the Web to smaller companies.[3]

On the logistics front, US industry alone spent an estimated $950 bn on logistics activities in 2000 and reportedly wants to cut costs in that area. But experts point out that the third-party logistics industry is lagging when it comes to e-technology and "must now speed up efforts to acquire technology that will enable it to electronically aggregate shipments, help customers visualize inventory levels, deploy transportation assets or save costs by consolidating shipments."[4]

There are exceptions, of course. AMR Research cites third-party logistics giants (3PLS) like APL Logistics, BAX Global, C.H. Robinson Worldwide, Exel, and TNT Logistics North America as examples of "e-enabled" companies. Companies like Descartes Systems and Freightquote.com are providing online logistics services. FreightMatrix offers an ASP (application service provider) model that provides virtual 3PL and supply chain services. And companies like DSC Logistics and Ingram Micro Logistic are now offering e-fulfillment.[5]

E-SOURCING/PROCUREMENT[6]

More and more, sourcing is becoming a global game. The Web opens up untold opportunities to locate the best goods at the best price from anywhere on earth. E-sourcing includes using Web-based technologies to automate and streamline the identification, evaluation, negotiation, and configuration of the optimal mix of suppliers, products, and services into a supply chain network that can rapidly respond to changing market demands, reports the Boston, Massachusetts-based Aberdeen Group. This emerging market includes "pure-play" e-sourcing solutions, reverse auction or dynamic trading technologies, supplier intelligence services, and tools as well as procurement service providers.

Speaking of procurement – what used to be called purchasing – the experts predict "staggering" benefits for companies and institutions with the capabilities to procure goods over the Web. E-procurement can provide everything from reduced transaction costs to reduction of stocks of obsolete products and inventory for those products, quicker reaction to market trends and redeployment of procurement and other professionals from tedious transaction processing to more strategic work.[7]

Although e-procurement is still in its infancy, a recent e-procurement study from the Center for Research in Electronic Commerce at the University of Texas of 1200 American and European companies indicates that a number of large companies are already reaping tremendous benefit from e-procurement. The study, funded by Dell Computer, looks at how manufacturing, retail and wholesale companies are conducting e-procurement. It addresses the effect that e-business "drivers" like system integration, supplier-related processes and supplier e-business readiness had on financial measures such as revenue/employee, gross margin, return-on-assets and return-on-investment.

COMMUNICATION (WIRELESS AND WEB CAPABILITIES)[8]

With the world of wireless and the Internet converging it won't be too many years before tech-savvy managers like you will conduct business from anywhere in the world using a Web-enabled phone. Although we're in the pioneering years of wireless global communication – and controversial years at that (see Chapter 6: The State of the Art) – there are still many developments taking place in both wireless voice and data transmission that will make managing a global operation far easier.

Web conferencing tools are available today that help you keep better contact with far-flung global teams. These include the ability to conduct "Webcasts" whereby information can be presented online in every format from PowerPoint to video.

Cellular technology, in particular, is evolving with the emergence of 3G (third-generation wireless) technology that allows for global roaming via cell phones. Two-way pagers – part of a so-called "follow me service" – are also starting to make an appearance. When your cell phone is turned off these devices send a message to your pager that alerts you to call your cell phone. The long-term aim is to consolidate all messages your receive and be able to reach you whether you're at home, on the road, or in a meeting, which would be a major boon to global logistics managers attempting to field messages coming from locations worldwide.

There are also evolving Web and wireless technologies that will allow managers to access e-mail via cell phones, for faster, more efficient information access while on the road. Even further down the road – and maybe just around the bend with the emergence of 3G wireless phone technology – will be the ability to put video on a cell phone. That will really cut through the maze of communication difficulties that confront any global logistics operation. But the experts warn not to expect that sort of miracle any time soon.[9]

IT'S SEEN AS A BOOST TO TRADE

Intelligent transportation systems (ITS) technology is already speeding cargo along highways in many countries worldwide. Besides providing pass-through capabilities at toll booths, the technology is employed to make pass-through highway safety checks for trucks. American, Canadian, and Mexican officials are experimenting with an array of ITS functions that could one day make crossing NAFTA borders both quicker and safer. For example, American officials are exploring new wireless applications that integrate GPS and ITS and allow relay of information on a vehicle long before arrival at the actual border crossing. The aim is to allow pre-clearance so drivers with clear records won't need to be inspected unless customs agents wish to perform a spot check.

On the Canadian border, the NAFTA Land Transportation Subcommittee is exploring opening up a corridor between New York and

Montreal to relieve trade crossings at Detroit, Buffalo, and Niagara Falls. ITS will be key to speeding up traffic while giving this effort a definite safety focus. Earlier efforts have been more on relieving congestion than on safety. ITS offers the opportunity to collect a great deal of data on shipments, carriers, and drivers.[10]

BEST PRACTICE
Telcordia promoting training

Telecommunications giant, Telcordia, is committed to building an environment in which effective leadership and management skills are practiced by everyone who has responsibilities for managing people. To achieve optimum business results, development of these skills is a priority. To this end, Telcordia is putting all 1,200 of its managers, including those abroad, through an intensive curriculum that includes five courses and is continuing to expand. The aim is to make "significant improvement in the way people were being supervised and evaluated, and make superior management a competitive advantage."[11]

E-transport and logistics in practice

SupplyLinks represents seven leading global transportation and logistics providers joining forces to offer procurement and transportation management services on the Web. Other SupplyLinks services include exception management across modes, carriers, and service levels to quickly identify delayed shipments and other in-transit issues. Another company plying its services on the Web is Mobility Technologies, a provider of digital traffic and logistics information. One product allows motorists in heavily congested areas to better plan routes and avoid congestion, which aids in just-in-time deliveries.[12]

Global computing

Sandvik Coromant Co. is a Swedish manufacturer of cutting tools with extensive North American and Latin American presence. To manage a global operation across an ocean and between two continents means being highly organized. There is no room for error. The company is already practicing a form of electronic remote logistics management utilizing three PCs with lots of computing power. This equipment stores and transmits data for the company's three distribution centers – two in Europe and one serving North America. Inventory tracking is managed from these centers, which then serve the global operation.[13]

Web-based conferencing

Nortel Networks is just one multinational cutting its travel expenses thanks to Web conferencing capabilities. Besides utilizing Net Meeting, a Microsoft tool, Nortel information technology experts have developed a proprietary tool called "meeting manager" to develop virtual meeting rooms. Virtual meetings take place this way:

Information is posted to a site and someone is selected to "'drive' the presentation. Everyone's at their desk and everyone goes through the issues, status of priorities, etc... Through an Internet browser you click on an URL. The tool allows you to actually point to the screen and draw on a slide while discussing its key points. Nortel is creating the infrastructure that makes this happen."[14]

GLOBAL BAR CODES WILL PROMOTE TRADE

The standards world is working on integrating computer hardware into bar code printers that will allow for integration with enterprise resource planning (ERP) business functions, and promote many new bar code functions to improve information flow throughout your global supply network. The International Organization for Standardization (ISO) subcommittee 31 on bar codes is working on standardizing two-dimensional (2D) bar codes. The advantage of 2D bar codes, whether they're stacked, matrix, or dot technology, is that they take up less space and typically have error-correction built in to enhance the read reliability.

When applied to a warehouse, 2D bar codes make it possible to encode more information on the package itself. Shippers can use 2D as a traveling database. Getting 2D bar code printing capabilities isn't hard.

All the major manufacturers offer it. But widespread implementation of these codes has only started in the last year or so on shipping labels to obtain data that would traditionally be on a packing sheet or slip.

Experts agree that the use of radio frequency identification (RFID) and bar codes with smart labels – basically, labels with embedded programmable chips – will grow quickly because this combination allows for long-distance identification and tracking of products. With RF technology you don't have to line up a laser beam to read the codes. And the chips can contain unlimited information that won't take up any more room on a label than traditional bar codes. Further, data contained in the chip can be edited.

The use? Let's say an order of commodity parts bound for one customer is re-routed at the last minute to another customer that needs the parts immediately, and will pay premium price to get them. At the distribution center, the smart labels can be edited not only to reflect the change in destination, but also to reflect the new delivery date and price. Moreover, these smart labels are durable. They can be baked into plastic or frozen. So a chip with RFID capabilities can be embedded on a frozen food container that's planted in a truck body to read information about the temperature and other conditions, allowing for a written history of the product during shipping.[15,16]

GLOBAL CODES SPUR PURCHASING

A new global classification code has been recently developed that should help you tech-savvy managers identify companies and classify products and services so that e-procurement can provide the efficient, cost-saving system of any manager's dreams. It's called the United Nations Standard Product and Services Classification Code (UN/SPSC) – a long name for what some consider one of the most useful coding systems ever devised.

The UN/SPSC is an internationally accepted eight-digit commodity code standard that classifies more than 8000 products and services worldwide. The first system, in fact, to classify both products and services for worldwide use, the UN/SPSC has been in operation for just about a year. It is based on the United Nations' Common Coding System (UNCCS) and Dun & Bradstreet's Standard Product and Services Codes (SPSC).

A number of companies have tried out UNSPSC, coding their catalogs to the global standard. The feedback from the front is that the standard is emerging, but immature. A Michigan-based electronics company found some product categories to be "not bad," but had difficulty with electronic components. The market, for example, differentiates between electronic and electrical switches, but UNSPSC does not. The good news is this same manager finds the UNSPSC process to be ever open and accommodating, and he is actively involved in improving the system.[17,18]

THREECORE HONES GLOBAL PROCUREMENT

You've been outsourcing logistics and accounting services for years. Why not try outsourcing your global sourcing of goods? That's the argument third-party procurement company ThreeCore Inc. has been making. The Massachusetts-based company specializes in finding, estimating, sourcing, and managing highly specialized parts for companies in areas such as engineered plastics, precision sheet metal, complex weldments (things you weld), machine parts (where it's a complex lathing operation), and printed circuit boards (PCBs).

Company officials point out that it makes sense to let specialists handle sourcing and procurement. Although the Web offers a wealth of new purchasing options, the leads generated are not screened and may provide more headaches and higher costs down the road. For example, goods from China may be far cheaper than those produced in New Jersey, but the shipping costs may make the transaction untenable.[19]

TRANSPORTATION.COM SETS UP FIREWALLS

While international organizations and governments grapple with standardizing e-commerce rules worldwide, companies like Transportation.com have developed their own strategies for dealing with online fraud, including establishing firewalls, tracing purchaser IDs, credit checks, and constant vigilance. The Kansas-based company provides online logistics services.

"We tend to take extreme measures – the same that would be practiced by a bank or a financial institution, although we don't have nearly that kind of exposure. But we still take that kind of protective measure. The industry has done well in that regard. Complacency is the biggest fear and physical security the biggest risk. Transportation.com surrounds itself with the appropriate firewalls and validates penetration around those firewalls."[20,21]

KEY LEARNING POINTS

» Advanced technology is a key factor in operating globally today. Technologies range from satellite to cellular, radio frequency, and IT.

» The Web creates an instant global dimension for any company or institution selling services, sourcing goods or services or buying online.

» New wireless capabilities will allow you to operate while on the fly.

» Intelligent transportation systems (ITS) are promoting pass-through capabilities on highways throughout the world and at borders.

NOTES

1 Zuckerman, A. (2001) "Speed 'n' Ease of Global E-comm Can Breed Legal Headaches." *World Trade*, September.

2 This material is adapted from Zuckerman, A. (2002) *Supply Chain Management*, Chapter 4: The E-Dimension. Capstone, Oxford.

3 Zuckerman, A. (2001) *Tech Trending*. Capstone, Oxford, p.116.

4 "Logistics Service Providers Fall Short in Web Enablement." *Supply Strategy*, May 2001 online edition. Sources are John Fontanella, service director, B2B marketplace service for AMR Research, Boston, Massachusetts, and Chris Newton, an analyst for AMR Research.

5 "Logistics Service Providers Fall Short in Web Enablement." *Supply Strategy*, May 2001 online edition.

6 This material is adapted from Zuckerman, A. (2002) *Supply Chain Management*, Chapter 4: The E-Dimension. Capstone, Oxford.

7 Fitzgerald, K.R. (2001) "Online Procurement of Production Materials Can Streamline Procurement Processes and Integrate Supply Chains." *Supply Strategy*, May.

8 This material is adapted from Zuckerman, A. (2002) *Operating Globally*, Chapter 4: The E-Dimension. Capstone, Oxford.

9 Material for this section is adapted from Zuckerman, A. (2002) *Operating Globally*, "Technology Boosts Communication Options for Global Logistics Operations". Capstone, Oxford. Source is Erik Bleyl, vice president of Transportation and Logistics, Aether Systems.

10 Zuckerman, A. (2001) *Tech Trending*. Capstone, Oxford, p.128 and Zuckerman, A. (2000) "Seamless Borders Nearing Reality," *Journal of Commerce online archives*, January 19.

11 Nicholas, G.P. "Telcordia's Management Training Offers Lessons for All Technology Companies." Blessing/White Inc. Source is Rosalind Doctor, executive director, Telcordia Learning Services.

12 Zuckerman, A. (2002) *Supply Chain Management*, Chapter 4: The E-Dimension. Capstone, Oxford.

13 Zuckerman, A. (2002) *Operating Globally*, Chapter 4: The E-Dimension. Capstone, Oxford. Source is Zuckerman, A. (1999) "It's Not a Small World After All." *Supply Chain Technology News*, September/October, pp.30–2.

14 Zuckerman, A. (2002) *Operating Globally*, "Technology Boosts Communication Options for Global Logistics Operations". Capstone, Oxford. Source is Patrick Sim, vice president of Nortel's Supply Chain Materials Management, Information Services.

15 Zuckerman, A. (2001) *Tech Trending*. Capstone, Oxford, p.106.

16 Zuckerman, A. (2000) "Belly Up to the Barcode." *Supply Chain Technology News*, September.

17 Zuckerman, A. (2001) *Tech Trending*. Capstone, Oxford, p.107.

18 Porter, A.M. (2000) "The Hunt for Interoperability." *Purchasing Magazine*, June 15.

19 Zuckerman, A. (2002) *Operating Globally*, Chapter 6: The State of the Art. Capstone, Oxford. Source is Joseph Tragert, vice president marketing, ThreeCore Inc.

20 Zuckerman, A. (2002) *Operating Globally*, Chapter 6: The State of the Art. Capstone, Oxford.

21 Zuckerman, A. (2001) "Someone Out There Wants to Cheat You!!" *World Trade*, March, pp.36–8. Rosalind McLymont conducted research for this article. Source is Daniel Bentzinger, senior vice president for information technology at Transportation.com.

The State of the Art

The evolving networked, Web-enabled world presents its own set of issues and concerns for tech-savvy managers to master. This chapter explores current trends in terms of how technology is affecting your operation. It includes:

» what is needed to make Internet/telecoms convergence work; and
» the legal and privacy issues that global e-commerce presents.

"If there's accord, then the party that hands over their IPRs then sets licensing fees that are based on "reasonable and non-discriminatory terms." The system is mostly self-policing, and most companies play by the rules. Industry wants an honor system and 99 percent of the time it works."

Amy Marasco, vice president and general counsel, American National Standards Institute (ANSI)[1]

From the infamous Microsoft anti-trust suit to concern over security in cyberspace, to the future of dot-coms and 3G wireless licensing disputes, the high technology world can be a contentious place. Even tech-savvy managers need to keep up with a whole slew of issues, not to mention incompatible software and equipment. One of your biggest chores will be keeping tabs on the numerous debates always taking place in some corner of the technology world. Here's a look at some emerging areas of controversy that could have an impact on your organization.

WHICH WAY WILL DATA FLOW?

The technology for transmitting data and information is undergoing a revolution. No one knows today what technology, or combination of technologies will reign as the supreme date/information transmission tool. And no one knows what combination of equipment or hardware will act as receivers. These are the sorts of questions that communication technology developers and service providers are all asking about the future of communication transmission:

» How will Internet/Web information be transmitted in five years – phone lines, wireless, TV cables, others?
» What hardware will be used to access this information – TV sets, phones, mobile phones, desktop and laptop computers, personal digital assistants (PDA), others?
» Where will you be able to access Internet/Web information – home, office, automobiles, airplanes, trains, gas stations, toll booths?
» Will microphotonics provide the enormous speed that its developers predict?

» Will bandwidth be able to keep up with the huge worldwide demand for information flow in cyberspace?

There is a second revolution underway that affects how you access information. Organizations still use paper to float information and content around offices and between trading partners, customers, and suppliers. But the paper-based world is rapidly being supplanted by electronic information flow. More and more information is being transmitted via the Internet, and in that enormous worldwide storage bin we call the Web.

Tech-savvy managers can take some comfort – or discomfort, depending on your bent – that even the technology experts don't know exactly what communications will look like in the next decade. Standards developers at the Geneva, Switzerland-based International Electrotechnical Commission (IEC) admit that they're a bit stymied about where to focus standards efforts in the computer, information technology (IT), communications arena.

CONSORTIA EMERGING AS A MAJOR FORCE

Myth may have it that new technologies emerge from the basements of struggling entrepreneurs. And there indeed are technology break-throughs, like the development of the operating system Linux, that can be attributed to individuals. But in reality, for the last 15 years or so, many standardized approaches to technology – and many of the new technologies that are promoting a Web-based world – are being developed in consortia such as RosettaNet, the World Wide Web Consortia (W3C), the WAP Forum, and many others.

These consortia allow members to lay their technology cards on the table in the form of intellectual property rights. Through sharing their intellectual property – a sort of high-tech copyright for computer code – they evolve new, more compatible technologies that are then applied to product design. Consortia also play a secondary role. Whether it's what cell phones the world will be buying next year, innovations in Internet security, how trading partners will communicate, or how cars will be networked for Internet access, consortia work can also give you a heads-up on technology's future course.

Most consortia are born and operate in the same way. Companies, sometimes with government assistance, choose a specific type of technology they want to make standard so it can be applied over and over without new research and development. The programming code for computer clocks, for instance, was hammered out in the Object Management Group (OMG) consortium. These groups have ranged from 2,500 active members (Internet Engineering Task Force (IETF), which supervises Internet maintenance) to as few as five.

A small group of committed companies and individuals can create compatible technologies that lead to a new generation of products. This is how the new international specifications for K56 modems were developed in 1998. Larger groups usually develop standards for existing, but conflicting, technologies. For example, The Open Group (TOG) has helped standardize Sun Microsystems' Unified Unix operating system that drives most of the Web.

Some consortia, such as RosettaNet, are even more ambitious. RosettaNet's 200 active individual and corporate members are developing over 100 specifications to develop integrated business-to-business communication. As part of this larger process, RosettaNet has developed what it calls Partner Interface Processes. PIPs, as they're known, will allow the computers of trading partners to communicate directly with each other. (See Chapter 2)

Retail giant L.L. Bean is conducting a pilot program utilizing PIPs. If this test is successful, the company has announced it will cease working in a paper-based system, or drop as much paper use as feasible today. Other companies exploring the use of PIPs are Arrow Electronics, CompUSA, Hewlett-Packard, Ingram Micro, Intel, SAP, and 3Com, all of whom have come out in RosettaNet press releases claiming at least some improvement in productivity based on early trial runs last winter.

Another large consortium, The Organization for the Advancement of Structured Information Standards (OASIS), is working on developing common web languages like SGML, SML, and XML that are crucial to the widespread use of e-commerce. Its major role is co-ordinating global standardization efforts for these languages to ensure worldwide compatibility between computer systems and web browsers. The consortium is also updating and making globally compatible the commonly used HTML, which is the basic language used to create Websites.[2,3,4]

CONSORTIA HAVE THEIR CRITICS

Consortia have their critics. Those from the traditional standards-developing arena, in particular, view consortia as interesting models for getting work done in a fast-pace, global economy. But they also view some consortia as closed systems that – with the wrong leadership – could end up being competition-busting cartels.

Traditional standards organizations also believe that consortia offer an inferior way of standardizing technology that doesn't involve the openness and accountability they can provide on a global basis. They believe that consortia duplicate work already taking place in the traditional standards world, which means expense for manufacturers and added costs for consumers. There are also fears that consortia can easily lead to the creation of cartels.

However, traditional standards setters also admire the ease and speed at which consortia function to standardize technology and meet market demands. They acknowledge that the traditional standards world had best learn from consortia if it wants to remain relevant to the high-tech industry. So for now, until another model emerges where technology can be standardized quickly and efficiently, this seems to be the era of the consortia.[5]

SOME SEE THE NEED FOR GLOBAL IPR RULES[6]

Intellectual Property Rights, or IPRs as they're commonly called, are central to the standardization of technology so crucial to interoperability and compatibility. Companies meet in consortia or traditional standards organizations and, when the system works as it should, openly share information on IPRs they hold on various patents. Most organizations have stringent rules guiding IPR exchanges, but most operate differently.

Standards experts and IPR attorneys who work globally encounter a variety of IPR rules worldwide. The discrepancies are confusing at best, and costly at worst. What to do about the situation is a matter of debate. On one end of the spectrum are a handful of standards experts who consider IPR rules as they exist today in many consortia and standards organizations "a nightmare," that when applied internationally often constitute non-tariff trade barriers. They predict many infringements of the rules and costly law suits unless industry focuses on developing stricter IPR rules and then harmonizes them on a global basis.[7]

But officials at high-tech trade associations and standards-related organizations such as ANSI argue that the rules generally work and are effective. Cases of malfeasance have been rare. One senior IT trade association officials says, "I can't make a sweeping case for reform based on isolated incidents, considering the substantial number of opportunities [for abuse of the system]."[8]

Other experts are concerned with potential abuse of IPR rules, but don't necessarily see either the need – or a viable way – of amending them. "No set of rules is ever going to be watertight, particularly when they rely on participants in the standards process acting in an open and fair or even ethical manner."[9]

Another common argument is that the system usually works and those who break the rules are "treated like pariahs, so I don't think you'll see much litigation. You have to be a pretty big company to have the war chest to have patents and set license fees, and it has to be a fairly unusual alignment of the stars before these things come up."[10]

There is some growing support, though, for addressing the question of whether harmonizing IPR rules on a global basis would help cut back on the potential for IPR to be manipulated as non-tariff trade barriers, and cut costs for industries selling product worldwide. A prominent consortium official says he "believes that organizations need to be able to understand what the differences in IPR rules are, and there should be more visibility of differences between rules. There should be a way of understanding IPR openly so they don't get caught. Right now, some companies or individuals unwittingly give up IPRs, or use standards in their products and then find a large bill coming later."[11]

MANY HURDLES NEED OVERCOMING TO OPERATE IN A WIRELESS WORLD

Whether it's the Internet and Intelligent Transportation Systems (ITS) or the Internet and telecommunications, international standards officials are pointing to the convergence between the Internet and telecommunications industries as this year's big issue.

Standards officials say convergence includes fixed networks and services. "The new idea is Internet telephony – VoIP (voice over IP). Mobility, in general, is growing dramatically. We're not just working on technical convergence, but convergence between the standards bodies that handle both telecoms and Internet so both worlds can co-exist. We're getting good results, especially in Study Group 13, Network Architecture."[12]

Although experts see great promise in convergence, they also believe a slew of issues have to be resolved before you will be surfing the Web on your mobile phone in airports worldwide or picking up your e-mail as you pass through toll booths. The following sections outline some of them.

Bandwidth issues and networks

For all the talk of wireless capabilities, we mostly function in a fixed wireless world today that's bound by fiber optic cables and telecom switches. As Internet/Web use soars, there's general talk of bandwidth crunches that could disable the entire cyberspace system.

» Standards developers – mainly from the telecom engineer side of life – are exploring a wide variety of ways to tackle network overload and avoid bandwidth crunches. Of particular concern is how to address transmission for voice – and increasingly data (including mainly multimedia communications) – over the "last mile" between the local exchange and the home or office. Standards are also being developed for the packet approach to network operations and transmission.

» "Forecasts indicate that there will be 250mn WAP (wireless Web) subscribers by 2003." That's on top of the 1bn subscribers who use fixed lines and "more than 1bn subscribers to mobile services. All of these developments signal more emphasis on the network and whether current bandwidth and switching capabilities can handle growth. All traffic needs to be covered through the network, whether that's fixed or based on radio or satellite [transmissions]. And we need to invest in switches."

» Other experts want to see "an agreement on how frequency bands are used globally. A lot of work is taking place within the ITU-R (International Telecommunication Union – Radiocommunication Bureau) to optimize the use of these bands. Much of this is political not technological. Different countries have different environments. Not everyone is using the same political, economic, or environmental approach. And the existing infrastructure and networks differ globally, particularly in the US, Japan, and Europe. That will matter for IP-based networks (e.g. the Internet)."[14]

Transmission modes for telecoms/Internet

ITU study groups are working on standards that will provide transmission systems for high-speed Internet access that will eventually replace the twisted pair cables for analog. Twisted pair technology – basically an old technology – may still hold some solutions today for bandwidth crunches. This is particularly crucial in the much-touted "last mile" where major costs to providers and users rest today.

Besides preparing new transmission systems using cables for high speed Internet that increase bandwidth over the existing twisted pair infrastructure – especially important for the local loop from the exchange to the provider – ITU study groups are also exploring packet switching, photonics, and other means of increasing bandwidth. Packet switching allows for data bytes transmitted through a telecom network to seek whatever route is available, rather than holding open one line for an indeterminate period of time. Telecoms experts believe this technology will free up vast amounts of line and help with increased volume. Photonics – and the newer microphotonics – are based on an "optical switch," which directly routes information in the form of photons of light without the need to convert to electronic signals, thus providing additional bandwidth savings.

Part of what makes this issue difficult is the fact that there are "any number of technologies that can be used to transmit data and information and no one knows which will predominate in the world of telecoms/Internet convergence. There's satellite, multi-point radio, cellular, and wireless LAN (local area network), for example."[15]

The future of 3G wireless in question

For several years 3G has been in the news. Not so long ago, this new generation of wireless technology was supposed to provide global roaming and all sorts of other wireless wonders. As has been well documented, European and Japanese telecoms companies have banked heavily on 3G and reportedly – particularly the Europeans – vastly over-spent on licensing fees for the technology.

ITU study groups have been prime developers of 3G standards. Now ITU officials see the future of 3G wireless literally up in the air just as the first release of 3G digital mobile communication specifications was announced at the end of May 2001. These specifications will allow manufacturers to deliver Third Generation Partnership Project (3GPP)-compliant equipment to operators "with the confidence that the specifications are comprehensive and stable," according to a release from the Alliance for Telecommunications Industry Solutions (ATIS), a major American-based telecoms trade association. The 3GPP represents global efforts to develop 3G standards and specifications.

ITU official Colin Langtry insists "there's a commitment to introduce 3G. Money was spent up-front to buy the spectrum and purchasers need a return on investment. We don't have 3G yet, but we will see the first systems rolling out this year. There are issues to resolve with 3G including frequency bands, global roaming. ITU is addressing these issues."[16]

On the radio side, he says ITU has developed "our first standard for the radio interface finalized. A World Radio conference has identified additional spectrum for this third generation of wireless standards – IMT-2000. The task is to optimize the use of this spectrum and facilitate global roaming. We're looking at the best ways to utilize frequencies to facilitate global roaming. And there are network issues. A call needs to be acknowledged and handled by a network."[17]

But other ITU officials agree with news reports that there are still many issues to resolve before 3G is a reality. For example, one can use a dualband phone in Europe, but we all must use a triband phone while traveling in Canada.

And they wonder whether in the drive to develop 3G that consumer interest and marketing have been lost. Do people really want to access the Internet in their car, for example?

US telecoms standards officials, while officially backing 3GPP, are becoming more dubious about the technology's future. They share ITU concerns that in the push to develop 3G standards – which are now officially published – business and individual consumer needs have been ignored.

One concern is that the screens on cell phones are just too small for adequate Web surfing. Another concern is that new wireless LAN (local area network) technologies – Bluetooth, intelligent transportation systems (ITS) and others – will make 3G obsolete. These technologies allow for short-range wireless communication in airports or at gas stations and form what one telecoms executive calls the "ethernet." They're reportedly less costly to produce than 3G and possibly even easier to use. Whether we'll see 3G products, the ethernet winning out, or the emergence of a rogue technology superseding all the others is anyone's guess.

Even so, wireless service providers like Nextel Communications in the US, and European and Japanese manufacturers, are all banking heavily on 3G. We can only wait and see what the next few years bring.

The VoIP and H.323 debate

Work is underway to promote voice over the Internet or VoIP, but which standards will predominate – meaning which technology may end up being marketed globally – is very much a matter of debate. The ITU has been promoting H.323, a standard for setting, modifying, and terminating telephone calls over the internet.

Other experts are touting SIP (Session Initiation Protocol). In a March 24, 2000 *Economist* quarterly report, they call H.323 a "Byzantine standard. There is no doubt that H.323 works, but it is needlessly complicated and understood only by the anointed few."

Senior ITU officials take umbrage at this and other similar reporting. They say the media is referring to a much earlier generation of H.323 and that the standard has evolved over the years, been simplified, and proved efficacious. "The media is comparing what was going on four or five years ago, but the standard has evolved and the specifications have completely changed," he says. "They should compare SIP with the latest versions of H.323. For example, the standard is about four times the size of what it was before and is more complex and future-oriented. At base, though, it's a very simple standard."[18]

Industry experts serving on ITU committees point out that the original H.323 was designed for the LAN environment and video conferencing. "After it was published the idea of sending voice over the Internet evolved. In fact, that was very shortly after its publication. People are making the wrong comparisons. They're looking for the version designed for video conferencing."[19]

But as is often the case in the world of technology and standards, there are market interests at stake in the development of a standard. Without naming names, ITU standards officials will say that "some people have a vested interest in not looking at how H.323 has evolved. These are a number of individuals, not necessarily companies."[20]

Moreover, they say that Study Group 16 isn't ignoring SIP. "We see it as an alternative to an important standard, but SIP doesn't handle multimedia right now. It's really designed for speech. Our aim is to provide interoperability between SIP and H.323. I wouldn't call it a merger. Some work has already been done in this area, but it will be a continuous process."[21]

No one on the standards side is offering any timetable for the day we'll be able to field phone calls via the Internet and hear voice via e-mail as we pass through toll booths, or gas up at a station. They counsel patience.

THE LEGAL ISSUES MULTIPLY WHEN YOU BUY/SELL GLOBALLY ON THE WEB[22,23]

It doesn't take much nowadays to peddle polka dot widgets worldwide. Design a Website, plug in an Internet connection, and you have global access to any market you can imagine, from Mozambique to Iceland, Malaysia to Brazil.

But what happens when those orders start pouring in? How are you going to guarantee payment? What if your product fails and you face an international law suit? How do you protect personal information like credit card numbers? What happens if the content on your Website is considered inappropriate or obscene in a foreign market?

The speed and ease of buying and selling globally on the Web can mean a bundle of legal headaches for those unfamiliar with the growing body of law that is emerging around global e-commerce. From electronic contracts and signatures to data privacy and security, the openness of the Internet/Web has created the need to offer legal protection to both global buyers and sellers that no one ever imagined even five years ago.

Until recently, cyberspace buying and selling – along with Web content – has been fairly unregulated on a global basis. But international lawyers with specialties in e-commerce law warn that those free-wheeling global e-trading days are over. Governments are starting to extend the reach of existing law to cover Internet commerce. Moreover, global e-commerce isn't the same as international trade in the three-dimensional world. Besides the ease of reaching foreign markets, the Internet creates legal issues that are unique to this medium.

Beware that when you buy and sell on the Web you will face privacy concerns, contract problems, and content may be restricted in some countries. Tech-savvy managers would do best to hire a good international lawyer and be prepared to pay him or her to keep them from incurring international disputes. Custom brokers and freight forwarders can also help you meet the rules and regulations that come along with global trade. And know that there are financial advisors – particularly international banks – that can help with letters of credit and other means of protecting your financial interests in a global marketplace.

KEY LEARNING POINTS

» The technology world is a contentious place that bears constant watching. Many debates are taking place relating to everything from security on the Web to intellectual property rights.

» Knowing which transmission pipe will provide your data is a major issue to tag in the future. Will it be cable TV, wireless, or fixed phone lines, for example?

» Much technology is either standardized or harmonized in high-tech consortia. These are organizations to tag and possibly join.

» Intellectual property rules provide protection for patented information and can also serve as non-tariff trade barriers. Many developments are taking place in this field that could affect your operation.

» One of the biggest trends that will affect you and your organization in the next five years is the convergence of the Internet with wireless telephone capabilities. For strategic advantage it's best to keep tabs on this arena.

NOTES

1 Zuckerman, A. (2001) "Rambus Case Raises Global Concerns." *Electronic News* online edition, April.

2 Zuckerman, A. (2001) *Tech Trending*. Capstone, Oxford, pp.180-5.

3 Zuckerman, A. (2000) "The Consortia Boom: Creating the Underpinnings of a Networked World." *Business 2.0*, October.

4 From a work in progress, *The Consortia Boom* by Amy Zuckerman.

5 Zuckerman, A. (2001) *Tech Trending*. Capstone, Oxford, pp.180-5 and Zuckerman, A. (2000) "The Consortia Boom: Creating the Underpinnings of a Networked World" *Business 2.0*, October, "Standards, Technology and Global Trends". Sources for both are Mike Smith, head of standards at ISO and Ronnie Amit, general secretary and CEO of IEC.

6 Zuckerman, A. (2001) "Rambus Case Raises Global Concerns." *Electronic News* online edition, April.

7 *Ibid.* Source is Carl Cargill, head of standards, Sun Microsystems.

8 *Ibid.* Source is Rhett Dawson, president of the Information Technology Industry Council (ITI).

9 *Ibid.* Source is Mike Smith, head of standards at the International Organization for Standardization (ISO).

10 *Ibid.* Source is Andy Updegrove, a partner in the Boston law firm of Lucash, Gesmer and Updegrove.

11 *Ibid.* Source is Allen Brown, president and CEO of the Menlo Park, California-based consortium The Open Group (TOG).

12 Source is F.J. Cantero, Head of the Promotion, Editing and Production department for the International Telecommunication Union's (ITU) Telecommunication Standardization Bureau (TSB) based in Geneva, Switzerland.

13 *Ibid.*

14 Source is Colin Langtry, counselor for the Study Group Department of the ITU Radiocommunication Bureau (ITU-R).

15 *Ibid.*

16 *Ibid.*

17 *Ibid.*

18 Source is Fabio Bigi, soon to be retired as deputy director for ITU's Telecommunication Standardization Bureau.

19 Source is Mike Buckley, rapporteur in ITU Study Group 16, whose focus is multimedia standards.

20 *Ibid.*

21 *Ibid.*

22 Zuckerman, A. (2002) *Operating Globally*, Chapter 6: The State of the Art. Capstone, Oxford.

23 Material for this section is adapted from Zuckerman, A. (2001) "Speed 'n' Ease of Global E-comm Can Breed Legal Headaches." *World Trade*, September. Rosalind McLymont helped research this article.

Success Stories

What are the secrets of managing a successful operation utilizing high-technology strategically? This chapter explains how ACUNIA, AvidXchange, Nextel Communications, and Seven-Eleven Japan are making it in the competitive high-tech arena. It includes:

» ACUNIA is driving the telematics industry;
» AvidXchange pushes MRO on the Web;
» Nextel promotes global roaming; and
» Seven-Eleven Japan wins with its own supply network.

For all the talk of the dot-com and high-tech meltdown there are many success stories taking place out there every day across the technology world. Far too many, in fact, to reproduce in this short space. The following companies and organizations were selected to offer a look at some of the innovative ways they are handling growth and/or management issues such as training.

ACUNIA IS DRIVING THE TELEMATICS INDUSTRY

"By integrating the ACUNIA Open Telematics Framework™ technology within Oracle's current product offering for in-car software, wired and wireless portals with CRM, we believe we are able to offer the automotive industry a working and future proof solution for the deployment of their future telematics and e-business management infrastructure."

Marc Maes, Co-founder and Co-CEO of ACUNIA NV[1]

A spin-off of IMEC, a Belgian microelectronics research company, ACUNIA designs telematics platforms for vehicles. Telematics involve a variety of technologies that, when combined, create intelligent navigation systems. ACUNIA product designs offer not only Internet connection for quick information download, but the ability to process information so that humans are removed from the link. Location and directions are provided without human intervention.

Founded in 1996 as SmartMove, ACUNIA is headquartered in Leuven, Belgium and, with an office in Frankfurt, Germany, the company employs about 120 people worldwide. The company's US subsidiary, ACUNIA Inc., has offices in Cambridge, Massachusetts and Detroit, Michigan.

In early 2001, General Motors Europe selected the company to develop its future telematics infrastructure and also announced partnerships and/or demonstration projects with, among others, Infineon Technologies, MW Group, Webraska, GenRad, and TeleAtlas. ACUNIA is well-known for its ACUNIA Open Telematics Framework™ (OTF), reportedly the first fully functional, end-to-end, Java-based open software design for the entire telematics pipeline. The single-platform, protocol-neutral, lightweight architecture system has won industry acclaim because it can be quickly, easily, and dynamically upgraded to add services and accommodate new technologies.

In July of the same year, ACUNIA NV and Oracle Corporation – one of the largest providers of software and hardware solutions for e-business – announced that they had formed a strategic alliance and will join forces to deliver integrated collaborative telematics-based m-Commerce solutions for the automotive industry. Oracle has chosen to promote the ACUNIA Open Telematics Framework™ technology within its Automotive Mobile Lifestyle solution, and jointly engage in the automotive industry in a bid to lead the next generation of service management and deployment technology.

According to spokesmen from both companies, the strategic alliance between Oracle and ACUNIA, which spans solution integration as well as joint sales and marketing, will enable the two companies to provide a comprehensive solution for enterprises within the automotive industry. Says an Oracle official: "There is a big challenge for the automotive industry to deliver the next generation telematics solution to their customers. The alliance between ACUNIA and Oracle brings a complete, scalable, and robust solution into being. Based on the latest products from Oracle, and adding the power of ACUNIA's OTF Server and components, the Automotive Mobile Lifestyle offering is the unique solution that enables a full solution from CRM to m-commerce."[2]

With company officials as young and energetic as its product lines, ACUNIA has been particularly smart in playing the PR game in both Europe and the US and in making strong contacts through standards organizations. Indeed, it was still a fledgling company as late as the first months of 2001 when it earned its first contract with GM, kicking business into high gear.

In fact, ACUNIA officials have been very active in developing telematics and intelligent transportation system (ITS) standards in the US. Company officials say standards efforts have paid off with solid public relations. Through its involvement in a number of standards organizations, the fledgling company has gained a growing reputation for being knowledgable and cutting edge. Clients in the US and the EU are signing up for ACUNIA technology, consulting, and design work.

ACUNIA works with several American standards developers, including the Automotive Multimedia Interface Collaborative (AMI-C), which is a unified forum for standardization of automotive multimedia technology. They've also donated time to the Open Services Gateway Initiative, whose concern is standardizing the transfer of data over wiring for home use for applications such as remote programming of washing machines, heat, or VCRs.[3]

Although at the time of writing it's too soon to determine how the ACUNIA/Oracle alliance will eventually work out, there's no doubt that ACUNIA is a company to watch. To move from ground zero as a European company and woo the biggest names in US automotive and high-tech is an accomplishment in itself.

ACUNIA TIMELINE

» **1996**: SmartMove founded in Leuven (near Brussels), Belgium.

» **1997**: IMEC, a major Microelectronics Research Institute in Europe, takes an equity stake in SmartMove as it begins to develop a communications standards program (DSRC) for the US Department of Transportation.

» **1999**: SmartMove selected to develop/participate in "Vehicle Telematics," an innovative traffic management plan in Flanders (Belgium). The program involves an architecture study, defining technical requirements, and building five prototype in-vehicle telematics communications systems for the Flemish government.

» **2000**:

  » February: SmartMove forms a US subsidiary and opens its first US operation in Boston (Cambridge), Massachusetts, that allows the company to draw upon top US software and hardware developers to support its growth strategy in the North American automotive and telecommunications markets.

  » May: SmartMove acquires Belgian software developer Information Technology Center (ITC), a specialist in the telecommunications industry.

  » December: General Motors Europe selects ACUNIA as telematics partner to develop third-generation telematics services for its European car lines; SmartMove changes its name to become ACUNIA.

» **2001**:
  » March: ACUNIA opens offices in Frankfurt and Detroit to maintain proximity to customers, making it easier to co-ordinate project management, telematics consulting, and sales activities.
  » July: Oracle Corporation, the largest provider of software for e-business, and ACUNIA NV announce that they have formed a strategic alliance and will join forces to deliver integrated collaborative telematics-based m-commerce solutions for the automotive industry.

KEY LEARNING POINTS

» In the fast-paced world of high-tech product development, having access to the best technology matters.
» Knowing how to play the standards game is crucial to success.
» Don't forget public relations. Name recognition counts.
» Determination and energy are also key factors, along with access to venture capital. It took ACUNIA five years to land a contract.

CISCO AND NORTEL PUSH OPTICAL INTERNET

Both Cisco Systems and Nortel Networks are promoting development and use of new technologies to promote more fiber optic capacity for Internet use. For example, Cisco is promoting Dense Wavelength-Division Multiplexing (DWDM), or multi-channel fiber optic technology, as an economical means of expanding existing fiber optic networks.

Reportedly, the Cisco DWDM platform can support up to 128 channels and reach 1.28 terabits per second of transmission capacity. It can be installed as a one-channel system and upgraded incrementally to its full capacity of 128 channels. The system includes an optical protection feature to help assure quality of service, and promoters say it has demonstrated its functionality and flexibility with a range of fiber types and channel combinations – critical issues for today's network providers.

Nortel Networks is also promoting technology to produce what they call "an all-optical Internet." It is building systems that will transmit data at speeds as high as 80 gigabits per second. Wider systems will enable 160 channels – or colors of light – to be transmitted on a single fiber no thicker than a human hair. The company claims that Nortel Networks Optical Internet systems can transmit light up to 4,000 kilometers without the need for electrical regeneration. Also in development are management systems that will allow service providers to manage individual channels of light, giving them the ability to quickly allocate new capacity to specific segments of the network or for specific applications on a network. Next-generation capabilities from Nortel Networks will provide much greater control and flexibility in manipulating individual channels of light inside DWDM networks.

Industry experts believe explosive bandwidth demand will push manufacturers to move into the terabit level, requiring an all-optical core. Today's "opaque" electrical core will be superseded by a "transparent," all-optical or photonic core that will offer the massive bandwidth, reliability, and performance available only from an all-optical network.[4]

AVIDXCHANGE PUSHES MRO ON THE WEB

"This new relationship between AvidXchange and Grainger provides us with everything we need to maintain our properties and offers us one-stop shopping for everything from ladders to grip tape. We now have a comprehensive list of Grainger products, a simplified general ledger and building coding, re-order templates from recurring purchases, electronic payment services and customized reports through AvidXchange's procurement system."

Daniel Levine, President, Levine Properties, Charlotte, North Carolina[5]

Headquartered in Charlotte, North Carolina, AvidXchange was formed in 2000 to provide Internet-based competitive bid management and purchasing tools and services to commercial and multi-family real estate companies. Through its software, AvidXchange enables real estate companies to create an improved, online method of buying products and services more effectively and cost-effectively from their suppliers and contractors. AvidXchange also provides catalog management, wireless communication, and online bill payment and presentment services for suppliers to increase sales and lower costs.

Within a year, this privately-held Web-based company had grown to 15 employees. In August of 2001, AvidXchange announced an alliance – what they called a "new business relationship" – with Grainger, one of the leading maintenance, repair, and operations (MRO) suppliers in North America.

Under this arrangement, AvidXchange's customers are given access to over 180,000 products to maintain and repair their facilities and equipment through Grainger's custom, online catalog. The catalog includes product description, pricing, and other critical information. The aim is to add speed and convenience to the purchasing process by making it easier for customers to locate and buy the products they need quickly and efficiently, according to company officials.

> "Grainger is an established entity in the MRO supply industry and can provide our clients with a large variety of products. This new relationship enables us to offer one-stop shopping for all our customers and provides Grainger with an added sales channel to reach the commercial real estate industry. We look forward to building this relationship with Grainger and providing our customers with even more value-added benefits."[6]

Besides the alliance with Grainger, AvidXchange provides a suite of Web-based procurement products that covers everything from electronic procurement to online bid management of service contracts and electronic payment of invoices. The company also maintains an alliance with consulting giant Deloitte & Touche to provide management and implementation of the Clarus eMarket technology.

AVIDXCHANGE TIMELINE

» **April 2000**: Founded.
» **August 2000**: Levine Properties completes bid for a 53,000 square foot project using AvidBid.

» **February 2001**: AvidXchange forms partnership with W.W. Grainger.
» **March 2001**:
  » AvidXchange expands to offer Supplier Express: e-commerce and marketing services to suppliers.
  » AvidXchange signs agreement with GE Capital Commercial Services to integrate financial services for commercial real estate owners.
  » AvidXchange forms partnership with XpedX.
» **April 2001**:
  » AvidXchange forms partnership with Hughes Supply.
  » AvidXchange forms partnership with Rent On The Dot.
» **May 2001**: AvidXchange Version 3.0 released.
» **June 2001**:
  » AvidXchange forms partnership with NOVA Lighting.
  » The Bissell Companies save $15,000 on a two-way radio bid using AvidBid.
» **July 2001**:
  » AvidXchange forms partnership with Management Reports International.
» **August 2001**: AvidXchange forms partnership with Essention.

KEY LEARNING POINTS

» E-commerce is ideally suited to the world of procurement, making it possible to access millions of buyers and sellers.
» Finding the right target product area where buyers and sellers can either cut costs or increase margins is key to success.
» The MRO arena, which is traditional purchasing turf, has been fertile ground for many companies venturing into e-commerce.

REINVENTING YOUR BUSINESS ON THE WEB

At Valley Books in Amherst, the owner sits behind the counter of his North Pleasant Street store chatting with walk-in customers and answering questions about his two floors of approximately

40,000 books. In between, he stays busy filling customer orders online, buying, and selling books over the Internet.

Fitness By Day is a Website that provides online personal fitness training. For $35 a month, members receive the benefits of "professional exercise program design and expert monitoring" by a certified personal trainer in the Boston area, who started the online program to fill downtime. Now his Website has testimonials from as far away as Santa Monica.

What these two businesses have in common is that they are not merely augmenting their traditional distribution channels: they are using the Internet to reinvent themselves. They are turning to e-commerce to expand their businesses by offering new, value-added services their customers want – services that were not even possible a few years ago.

Consider how travel agencies are reinventing themselves to operate on the Web. Airline ticket buyers are finding it easier to go online and view screen displays than to try to jot down times and flight numbers over the phone. The Web also allows customers to play with scenarios at their own pace on their own time, find bargains, and bid on tickets. In 1999, 16mn travel reservations were booked on the Web. Travel agencies must learn to provide new, value-added services or they will perish. Is your business similarly imperiled?

Reinvent yourself!

There is hardly a business in the "New Economy" that can perform without some Internet connections. This is true whether your company is Internet-based, like eBay or Amazon, or Internet-enabled, meaning it uses the Internet to support the business, but is not 100% dependent on it. Either way, success today will hang, to some degree, on how you work the Web.[7]

NEXTEL PROMOTES GLOBAL ROAMING

"Nextel is one of the most innovative companies in the US. They were among the first, if not the first, to introduce no roaming charges, one-second billing, no long distance charges and packet-based wireless Web services. They were also the first company to really offer global roaming to customers.''

Jeffery Hines, managing director and group head, North American Telecommunications Research Deutsche Bank[8]

Since its beginnings in 1987 as Fleet Call, Inc., Nextel Communications has become one of the top six cell phone service providers in the United States, with a growing reputation worldwide as one of the few service providers that offer global roaming options. A leader in providing fully integrated, all-digital wireless service, Nextel has approximately 16,000 employees and is working on developing a nationwide CMDA network to be officially operative in 2002.

As of 2000, Nextel posted revenues at $5.7bn and in 2001 listed 7.2mn domestic digital subscribers and 8.3mn global digital proportionate subscribers. It maintained international offices in Argentina, Brazil, Mexico, Peru, and the Philippines and had investments in wireless companies operating in Canada and Japan.

Nextel, whose slogan is ''Go Global with Nextel Worldwide'' and which boasts cell phone coverage in 85 countries worldwide, has carved out a niche focusing largely on the business customer. Besides the capability to provide the closest thing to global roaming on the market, Nextel was reportedly the first company to introduce Java to wireless. Java is a platform that allows the phones to run a variety of applications, from games to specific functions for business such as time sheets.

Nextel officials say the company competes on product and services, not price. ''We're the only ones that offer services like data on a single hand set. We grab 11% of new wireless customers every month.''[9]

Despite its business emphasis, Nextel was pushing into other markets in 2001. Analysts point out that through their indirect channels of distribution they have sold their product to a much wider audience. Also, a recent acquisition of the Let's Talk Wireless stores is supposed to increase Nextel market presence in less traditional business segments for wireless. Basically, analysts argue that Nextel has no choice but to broaden its scope. Their near monopoly on the business sector can't last and the wireless giant will have to become a full-fledged competitor across all markets to stay competitive in the future.[10]

Nextel is looking to the next generation of wireless technology – the so-called third generation or 3G – to propel them in the marketplace. The new technology boasts global roaming capabilities and superior quality in voice. There's a firm belief among Nextel management that the world is ready to embrace the freedom that wireless communication offers. And they believe that their largely business customer base is eager to see the roll-out of technology that will combine telecoms functions with computing and the Internet. Although this sort of equipment is still in the design phase and standards are just starting to be hammered out to provide global interoperability, they point to the popularity of Nextel's packet networks that provide data capabilities along with voice as evidence for the business public's love affair with wireless.

NEXTEL TIMELINE

- » **April 1987**: Fleet Call, Inc., Nextel's predecessor, is founded.
- » **March 1993**: Fleet Call, Inc. changes its name to Nextel Communications, Inc.
- » **July 1994**: Nextel and OneComm, formerly CenCall Communications, Inc., announce a merger, forming a company to provide service in all of the top 50 US markets.
- » **August 1994**: Nextel agrees to buy all of Motorola's SMR radio licenses in the United States, providing Nextel with significant spectrum rights in each of the top 50 US markets.
- » **January 1997**: Nextel introduces the Nextel National Network and announces it will not charge roaming fees for customers traveling anywhere on its digital network – a first in the wireless industry.
- » **March 1997**: Nextel announces a new pricing program, including rounding calls to the nearest second after the first minute, and launches its first national advertising campaign with the tagline "Get Smart, Get NextelSM." iDEN service was also launched in Northern California.
- » **September 1997**: McCaw International, Ltd., Nextel Communication's wholly-owned subsidiary, changes its name to Nextel International, Inc. The newly-named company is a leading international wireless communications services company with wireless operations and investments in Canada, Mexico, Brazil, the Philippines, Indonesia, Argentina, Peru, Japan, and Shanghai.

» **August 1998**: Nextel Communications Inc., along with other equity investors including Eagle River Investments, Motorola, and affiliates of DLJ Merchant Banking Partners II (an affiliate of Donald, Lufkin & Jenrette), makes an agreement with Nextel Partners, Inc. to build and operate an integrated digital wireless communications network utilizing frequencies licensed to Nextel, the Nextel name and Motorola's iDEN technology in mid-size and small markets throughout the United States that are not presently built out or are in early stages of build-out by Nextel.

» **February 1999**: Nextel announces Nextel wireless data services – a family of wireless Internet services designed to combine and leverage the power of the Internet and the Nextel National Network to bring the integration of wireless voice, data and messaging to Nextel customers.

» **June 1999**: Nextel introduces the i1000plus™ Internet-ready phone. Reaches agreement with the US Department of Justice for an out of court settlement regarding Nextel's legal challenge to the 1995 Consent Decree. McCaw family announces plan to exercise options to purchase 17mn Nextel shares.

» **February 2000**: Nextel announces alliances with Aether Systems, Amazon.com, eDispatch.com, and IBM to offer wireless Internet applications unique for businesses through Nextel OnlineSM.

» **April 2000**: Nextel announces Nextel OnlineSM, the wireless Intervnet solution for business: an "always connected" wireless data solution that provides business customers with the tools and applications they need to access time-sensitive content and information instantly to get business done. This service is launched in 43 major markets, including more than 750 cities across the US.

» **April 2000**: Nextel announces Nextel WorldwideSM, the largest all-digital wireless coverage in the United States and in more than 70 countries around the world including 18 of the world's 25 largest cities. With the i2000™ phone, the first and only dual mode iDEN®/GSM 100% digital cellular phone, customers can now use the same phone number no matter where they are, whether it's across town, in another country, or around the world.

» **June 2000**: Nextel launches Nextel Online Two-Way MessagingSM service that enables customers to send, receive, and respond to text messages using their Internet-capable Nextel "plus" phones.
» **November 2000**: Nextel gains access to additional 800 MHz band spectrum.
» **April 2001**: Nextel introduces First Wireless Java™ phone in North America.

KEY LEARNING POINTS
» Focusing on a market niche can pay off.
» Producing new and dynamic technologies and promoting them to hell is a winning strategy.
» Moving into the international arena is the pathway to future success.
» Staying flexible and nimble is a necessity for a large corporation.

SEVEN-ELEVEN JAPAN WINS WITH ITS OWN SUPPLY NETWORK

"Seven-Eleven's merchandising and product-development capabilities are formidable. Its ability to sense new trends and churn out high-quality items is far superior to other operators."
Michael Jacobs, analyst, Dresdner Kleinwort Wasserstein[11]

Seven-Eleven is best known as an American convenience store chain whose stores go under the name 7-Eleven. Since 1973, though, the corporation has maintained a Japanese subsidiary – Seven-Eleven Japan. Seven-Eleven Japan operates more than 8300 7-Eleven stores in Japan and Hawaii under an area license agreement with 7-Eleven.[12]

Thanks to innovative uses of technology that amount to a proprietary supply network, Seven-Eleven Japan has made huge strides in that country. In fact, the *Economist* reports that Seven-Eleven Japan in May of 2001 "snatched the title of biggest retailer in Japan from Daiei, a troubled supermarket giant."[13]

Until recently, the company's e-strategy depended mostly on proprietary solutions. It has largely used the Internet to communicate with customers rather than handling business functions. But the company has earned praise for the way it uses technology, particularly its pioneering use of the Internet and other cutting-edge technologies.

In fact, the company has a long history of using technology in innovative ways. In the 1980s, the company replaced cash registers with point-of-sale (POS) systems that monitor customer purchases. A big fan of information technology, it had installed a number of systems and upgrades by 1992. In 1995, just as the Internet was taking hold in the US, Seven-Eleven Japan installed proprietary technology to promote IT work rather than moving information to the Net.[14]

The demand, at the time, was for a system that could handle multimedia pictures and sound, that was user friendly and easy to maintain. The company also wanted a system that would speed up transmission of orders, could field ideas, and could provide feedback throughout the chain, and it wanted everyone in the supply chain on one system that was easily maintained and updated, and would last 15 years. Basically, it was designing a supply network without calling it that.

All of this work was taking place as enterprise resource planning (ERP) software was just starting to emerge in the US and Europe, along with the Internet. However, ERP in those days could only handle back-office business functions for one business and Seven-Eleven Japan wanted to network 6000 stores, a number that has swelled to 8500 in the intervening years.

The company hired NEC, a consumer electronics company, to design this system. Eventually, they tapped Microsoft to develop software for them to work on a Windows-based system. The software was ready for downloading to some 61,000 computers at Seven-Eleven's stores by 1996. In 1998, the company overhauled the system at a cost of hundreds of millions of dollars, but included more interoperability and direct support from its base in Seattle, Washington in the US. For transmission, the company depends on satellite technology, which is cheaper than ground cables and can withstand earthquakes. Also, fixed-wire options aren't available in some rural areas.[15]

Despite the cost and being somewhat cumbersome by today's standards, Seven-Eleven Japan's supply network has provided a multitude of advantages. It's capable of monitoring customer needs and promotes a "pull" supply chain system where customer demand drives orders and manufacturing, rather than the "push" system of not long ago where product was created to be promoted and sold with much wasted inventory. Sales data and software, which collect information from all stores, help promote quality control, pricing, and product development so that customer needs and interests can be quickly analyzed and changes made to meet demand.

Moreover, the technology has helped Seven-Eleven Japan move to daily trends and cut back on long-term forecasting. And it's made the company's supply chain more efficient, including providing for faster orders, better inventory management, and quicker shipments. The system has also aided Seven-Eleven vendors and manufacturers with inventory management, helped on the employee training front, and improved communication throughout the organization.

Now five years into operating on a supply network, Seven-Eleven Japan is assessing how much of the system to move to the Internet/Web. The company has increased customer traffic by turning shops into payment and pick-up points for Internet shoppers. But security and payment are still major concerns. In a country where credit payment is discouraged, reportedly 75 percent of Seven-Eleven Japan customers still pick up their purchases at bricks-n-mortar-stores rather than rely on shipment. Despite these sort of cultural obstacles, Seven-Eleven Japan has purchased and plans to install an e-commerce software package to push e-procurement for bulk goods and services such as office equipment and insurance policies for employees. Moving to online sales and promoting online brands is under scrutiny.[16]

SEVEN-ELEVEN JAPAN TIMELINE

» **1927**: 7-Eleven, Inc., then the Southland Corporation, is founded in Dallas, Texas, pioneering the convenience store concept during its first years as an ice company with retail outlets.

» **1946**: The name 7-Eleven originates when the stores open from 7 a.m. to 11 p.m.

» **1973**: The company opens its first stores in Japan under Seven-Eleven Japan.

» **1991**: IYG Holding Company, a wholly-owned subsidiary of Ito-Yokado Co., Ltd., and Seven-Eleven Japan Co., Ltd., takes a majority interest in 7-Eleven, Inc.
» **1999**: Its corporate name is changed from The Southland Corporation after approval by shareholders on April 28.
» **2000**: Through charitable contributions of cash and goods, in-store fundraising activities and local involvement, more than $2.3 mn in support is disbursed to programs addressing issues such as literacy and reading, crime, and multicultural understanding.
» **2001**: The company operates 21,000 units worldwide.

KEY LEARNING POINTS

» Tailor technology to a culture and the terrain.
» Proprietary technology can work, though it can be costly up-front.
» Innovative thinking and the willingness to make change, plus pocket the price, can pay off.
» Jumping to the latest technology is not always the wisest move.
» Developing a supply network may be costly in the short haul, but promotes cost efficiencies and improved communication that almost always reap huge benefits.

MOTOROLA: REMOVING THE WEAK LINKS

Identifying the weak links in a supply chain is essential if tech-savvy managers are to develop supply networks and other collaborative cyberspace arrangements. As simple as it sounds, soliciting supplier feedback is a major key to analyzing what is and is not working. But many companies and institutions don't bother.

That's not the case at Motorola, where ensuring a two-way flow of information has been formalized in a one-page checklist with about 20 traits, including early supplier involvement, response to cost reduction ideas, purchasing professionalism, forecast accuracy, and others. The supplier rates Motorola on a six-point scale against each trait. The supplier is asked to rate its best customer, whose identity is not revealed.

Motorola also uses activity-based costing to assess its supply chain. The challenge is to identify cost drivers that affect the supply chain. Some of the criteria to establish the overall ratings include lead time, cost, current supplier, sole versus single source, and custom versus standard component. Another Motorola trick is establishing a supply chain ombudsman to "arbitrate" between the customer and supplier. Finding ways of sorting out disputes, assessing the "truth," and getting both parties to cooperate is part of this job. And an astute ombudsman will find opportunities to expose weak links in the supply chain and means of strengthening them.

When the semiconductor giant outsources work, it ensures that the service provider is viewed as an extension of the internal factor. That means sharing planning, inventory, human resources, and information technology systems, and keeping different corporate cultures in mind.[17]

NESTLÉ GOES WEB-CENTRIC

Some ERP vendors are staying alive by selling supply chain management applications. Others are moving beyond the supply chain to the Web. Enterprise software giant SAP made recent headlines signing food giant Nestlé SA to its largest software deal ever: $200 mn for mySAP.com Web-based applications that will be accessible to all of Nestlé's 230,000 employees in 500 factory sites worldwide.

Nestlé plans to overhaul its entire core business on the Web-based software platform, which will function as role-based and personalized enterprise portals for the Swiss candy maker's entire staff. Nestlé employees will have access to internal and external applications in such areas as supply chain management, product manufacturing, customer relationships, knowledge management, e-commerce, product lifecycles, financial and cost management, and marketing. SAP will customize the portals so that only those employees whose jobs are defined by a specific role can gain access to information required by their jobs.

In other attempts to keep pace with competitor Oracle Corp., SAP also purchased a $250 mn stake in e-procurement solution provider Commerce One (www.commerceone.com). Not only does this pact give SAP a sizable share in one of the leading e-marketplace makers, but it also provided SAP with an entry point into the Covisint auto exchange being formed by the Big Three. Commerce One and Oracle are both providing the technology that will power this online exchange. SAP has announced plans to bring Covisint beyond the US and link it to global markets.[18,19]

NOTES

1 Source is an ACUNIA press release, July 24, 2001.

2 Source is Franck Boutboul, vice president of Industrials & Consumers, Oracle EMEA.

3 Zuckerman, A. (2001) *Tech Trending*. Capstone, Oxford, p.165. Sources are Stephen Buytaert, CEO of SmartMove and Bart Stevens, American marketing director for SmartMove.

4 Zuckerman, A. (2001) *Tech Trending*. Capstone, Oxford, pp.32–3. Sources are press releases from both Cisco Systems, San Jose, California, and Nortel Networks, Branpton, Ontario, summer 2000; Greg Mumford, president, Optical Networks (a division of Nortel Networks), and Dale Misczynski, president of The ISOagile Group, Austin, Texas.

5 Source is a press release from AvidXchange, August 8, 2001.

6 Source is Mike Praeger, CEO of AvidXchange, from an August 8, 2001 press release.

7 Zuckerman, A. (2001) *Tech Trending*. Capstone, Oxford, p.98. Source is Goldberg, E. (2001) "Taking it to the Web." April 21, info@entrenetwork.com. Information source is Les Ball, co-director of the Interdisciplinary Center for Electronic Enterprise (ICEE) at the University of Massachusetts Computer Science Department.

8 This quote appears in an article for *Continental* by Amy Zuckerman on Tim Donahue, CEO of Nextel Communications. Publication is scheduled for December 2001 or January 2002. No headline has been set.

9 Source is Tim Donahue, Nextel president and CEO.

10 Source is Jeffrey Hines.

11 "Over the Counter E-commerce: How to blend e-commerce with traditional retailing." *The Economist*, May 26, (2001), p.77.

12 Source is www.7-eleven.com.

13 "Over the Counter E-commerce: How to blend e-commerce with traditional retailing." *The Economist*, May 26, (2001), p.77.

14 *Ibid.*

15 *Ibid.*

16 *Ibid.*, p.78.

17 Zuckerman, A. (2001) *Tech Trending*. Capstone, Oxford, p.241. Source is Mazel, J. (2000) "Remove Supply Chain's Weak Links No Matter Whose Fault They Are." *Supplier Selection & Management Report* July 1. Source is Jim Limperis, senior contracts manager, Supply Management, Motorola, Inc.

18 Zuckerman, A. (2001) *Tech Trending*. Capstone, Oxford, p.243.

19 Blanchard, D. (2000) "SAP's Sweet Deal." *Supply Chain Technology News*, August 14, p.1.

Key Thinkers/Concepts

Technology has its own language. Get to grips with the lexicon of technology, and the key concepts, with the ExpressExec technology glossary in this chapter. It includes:

» the top technology entrepreneurs and developers of the post-war era;
» many of the most important technology developments since the 1940s and their impact on your operation; and
» a glossary of common technology terms.

It's not hyperbolic to suggest that hundreds of thousands of people worldwide have contributed to developments in high-technology that have produced the networked, Web-enabled world that tech-savvy managers work in today. There is space here to include only a handful of key thinkers and players who have had made major contributions to the evolution of Information Technology as we know it today. Unlike other fields, where those who work on the theoretical side are often divorced from the "real world," the IT arena has blossomed because so many key players bring their IT knowledge to bear in an entrepreneurial fashion. The following is a sample of those thinkers and entrepreneurs whose work has shaped our technology universe.

MARC ANDREESEN – INTERNET/WEB PIONEER AND ENTREPRENEUR

Known as the founder of the online portal, Netscape, Marc Andreesen has contributed more to the world of technology than the development of a business alone. In 1992, Andreesen and Eric Brina unveiled Mosaic, an early Web browser, which helped bring the Internet and Web into the homes of millions of users and made both men Web pioneers. Mosaic was more sophisticated graphically than other browsers of the time, allowing images and text to appear on Web pages and including for the first time the hyperlink to retrieve documents from other sites.

In 1994, after moving to the Silicon Valley in an effort to mass-produce Mosaic, he and Jim Clark formed Mosaic Communications Group, from which Netscape would evolve. On October 13, 1994, what was then called Mosaic Netscape was posted for download on the Internet and soon became the browser of choice for millions of users. Netscape was sold to America Online in 1999 for $10 bn in stock.[1]

DAN BRICKLIN – PC COMPUTING PIONEER AND ENTREPRENEUR

Dan Bricklin is best known as developer of VisiCalc, one of the first serious business applications that did for numbers what word processing did for words – enabled the user to insert and delete elements and see an immediate change in the results. Bricklin, a graduate of the Massachusetts Institute of Technology (MIT), developed

VisiCalc while at Harvard Business School, which he had entered after several years working for Digital Equipment Corporation (DEC) in the mid-1970s. Together with Bob Frankston, a fellow MIT student, and Dan Flystra, owner of Personal Software, they refined VisiCalc and prepared it for mass marketing. Eventually, it was used in both the Apple II and Tandy TRS-80 computers, and is credited with boosting the fledgling Apple Computer company into a major market contender as many business people bought Apples just so they could run VisiCalc.[2]

TIMOTHY BERNERS-LEE – DEVELOPER OF THE WORLD WIDE WEB

If tech-savvy managers are relying more and more on that grand cyberspace storage bin called the World Wide Web, you can partially thank an English computer scientist named Timothy Berners-Lee. The Web's beginnings date back to circa 1980 when, fresh from Oxford, Berners-Lee wrote a software program called Enquire while working at the European Laboratory for Particle Physics (formerly known by the acronym CERN) near Geneva, Switzerland. Originally intended to help him remember connections between various people in his lab and to connect his work to their projects, Enquire became the mental roadmap to a far more grandiose idea of developing a worldwide electronic linking system.

Eventually he and CERN scientists married an information system that would create a web of information to the Internet and introduced this concept – what became the World Wide Web – in 1989. Their fledgling effort subsequently became a platform for related software development, and the numbers of linked computers and users grew rapidly to support a variety of endeavors, including a large business marketplace. Its further development is guided by the WWW Consortium based at the Massachusetts Institute of Technology in Cambridge, Massachusetts.[3]

STEVE CASE – WEB PORTAL PIONEER AND ENTREPRENEUR

Today America Online (AOL) is a household name, the company is so large that it recently bought out Time Warner, and the concept of using a portal to enter the World Wide Web is old hat. But no one had heard of the concept in the mid-1980s when an unknown 25-year-old named Steve Case crafted the concept of Web portals and unveiled a company originally called Online America. Case had a background in political science and marketing, not computer science. It was through a job with Control Video in 1983 that he entered the world of technology and video games and was onboard, trying to keep the company afloat, when in 1985 it began to provide online services for Commodore computer users. Renamed as Quantum Computer Services, the business eventually became Online America and then America Online.

In 1992, with just 120 employees, AOL went public, raised $66 mn and Case was made CEO. AOL bought out longtime rival CompuServe and launched its famous mass-marketing drive by sending millions of disks to households and attaching them to magazines. It then bought Netscape and formed an alliance with Sun Microsystems. In 2000, AOL bought media behemoth Time Warner for $166 bn, and became a full-blown media company.[4]

DOUGLAS ENGLEBART – SOFTWARE PIONEER

Today we all take the computer mouse for granted, along with the concept of windows and many other common PC functions. But the world has Douglas Englebart to thank for many of the tools that tech-savvy managers use each day. He started his work developing everything from the computer mouse, windows, shared-screen tele-conferencing, hypermedia, groupware, and more, in the late 1960s. He unveiled a number of his memorable patents at the Fall Joint Computer Conference in San Francisco in 1968. Many of the inventions, particularly the mouse, used for navigating PC programs outside of the keyboard, were ahead of their time. For example, it wasn't until 1984 that that the Apple Macintosh popularized the mouse. Engelbart now works out of the Bootstrap Institute, which he founded, where he is an inventor and a consultant in multiple-user business computing. His current focus is on a type of groupware called a "open hyper document system," which may one day replace paper record-keeping entirely.[5]

WILLIAM GATES – SOFTWARE PIONEER AND ENTREPRENEUR

In America, the 1990s and the early years of the new century were dominated by two Bills – Bill Clinton, then president of the US, and Bill Gates, chairman of the formidable Microsoft Corporation. Started in Seattle by Gates and colleagues in the mid-1970s, Microsoft has become one of the largest, most powerful and controversial companies in the world, with revenues of $22.9 bn in the fiscal year ending June 2000 and with 39,000 employees in 60 countries.

Gates has spent virtually his entire career in software and with Microsoft. He attended Harvard University in the mid-1970s, where he met Ballmer, now Microsoft's chief executive officer. While at Harvard, Gates developed a version of the programming language BASIC for the first microcomputer – the MITS Altair. In his junior year, Gates left Harvard to devote his energies to Microsoft, a company he had begun in 1975 with his childhood friend Paul Allen. Guided by a belief that the computer would be a valuable tool on every office desktop and in every home, they began developing software for personal computers.

Under Gates' leadership, Microsoft's mission has been to continually advance and improve software technology, and to make it easier, more cost-effective, and more enjoyable for people to use computers. The company is committed to a long-term view, reflected in its investment of more than $4 bn on research and development in the current fiscal year. However, Gates and his company have also faced a slew of law suits – including a monumental anti-trust suit waged by competitors via the US Justice Department – that allege that Microsoft has conducted business in a monopolistic fashion. A devoted philanthropist, his Bill and Melinda Gates Foundation has donated billions of dollars to support the use of technology in education and skills development.[6]

GRACE MURRAY HOPPER (1906–1992) – SOFTWARE PIONEER

Computer programmers rely on COBOL, the first programming language user-friendly to business, but few people realize that the inventor was Grace Murray Hopper, who worked both in academia and in the Navy, from which she retired as a Rear Admiral in 1986. In a life dedicated to proving that the impossible is possible, Hopper invented the first computer "compiler" in 1952. This revolutionary software facilitated the first automatic programming of computer language. Before Hopper's invention, programmers had to write lengthy instructions in binary code (the base "language" of computers) for every new piece of software.[7]

STEVE PAUL JOBS – COMPUTER PIONEER AND ENTREPRENEUR

Known in his early years as the "bad boy" of the computer world, Steve Jobs, along with Steve Wozniak (see below), co-founded Apple Computers Inc. Introducing their first computer, the Apple I, in 1976, they helped launch the world of personal computers that so many of you tech-savvy managers rely on today. From the Apple I came the concept of the mouse for easy navigation and the graphic display concept that Microsoft evolved into Windows. Going public in 1980, the company also released the Apple II and III and eventually developed the popular Macintosh, designed to compete with the PC, which was released in 1984.

Despite his technological brilliance, Jobs had difficulties with managers like John Sculley, whom he brought in to run Apple. When major layoffs hit in 1985, he lost a power struggle with Apple top management, quit the company, and formed the NeXT corporation. Ironically, Jobs eventually re-joined Apple in the late 1990s where he has been developing his own line of products including the iMac.[8]

JACK KILBY – ELECTRONICS PIONEER

Nobel Prize recipient Jack Kilby, an engineer with Texas Instruments, is considered one of the greatest electrical engineers of all times. In 1958, he invented the integrated circuit, which combined three electronic components onto a small silicon disc, then named the microchip. It is from this simple quartz object that modern computing derives its roots. The microchip made microprocessors possible, and therefore allowed high-speed computing and communications systems to become efficient, convenient, affordable, and ubiquitous.

All told, Kilby has more than 60 patents to his credit. He went on to develop the first industrial, commercial, and military applications for his integrated circuits – including the first pocket calculator (the "Pocketronic") and computer that used them. An independent inventor and consultant since 1970, Kilby has used his own success to promote other engineers and inventors – most notably by establishing the Kilby Awards Foundation, which annually honors individuals outstanding in science, technology, and education. In 2000, he was awarded the Nobel Prize for Physics for his part in the invention of the integrated circuit.[9]

LEONARD KLEINROCK, PHD – INTERNET PIONEER, ACADEMIC, AND ENTREPRENEUR

Leonard Kleinrock is known as the inventor of Internet technology. In the late 1960s, while a graduate student at MIT, he created the basic principles of packet switching, the technology underpinning the Internet. Moving onto UCLA as a professor of Computer Science, his Host computer became the first node of the Internet in September 1969. He wrote the first paper and published the first book on the subject and also directed the transmission of the first message ever to pass over the Internet.

Recipient of many awards and honorary degrees, he is a co-founder of Linkabit. He is also Founder and Chairman of Nomadix, Inc and of Technology Transfer Institute, both high-tech firms located in Santa Monica, California. Additionally, Kleinrock has recently launched the field of nomadic computing, the emerging technology to support users as soon as they leave their desktop environments; nomadic computing may well be the next major wave of the Internet.[10]

LINUS TORVALDS – OPERATING SYSTEM PIONEER

Linus Torvalds, a young, entrepreneurial software designer from Finland, has helped launch the open source movement through development of his software operating system, Linux. By openly sharing Linux with the world, rather than selling the software in the fashion of Microsoft, he has rocked the software industry and created enormous challenges for companies like Microsoft. Torvalds developed the first version of Linux while a computer science student at the University of Helsinki. By October, 1991, Linux 0.02 was announced to the world. In 2000, through the hard work of Linus and many other people, a more user friendly version of Linux – 2.20 – was introduced and rapidly became an extremely useful and popular operating system.[11]

AN WANG, PHD (1920–1990) – IT PIONEER AND ENTREPRENEUR

A Shanghai native, An Wang played an enormous role in the evolution of post-war electronics and computing in the US. Inventor of the Pulse Transfer Controlling Device in 1955, An Wang made magnetic core memory a practical reality. Eleven years later, he introduced LOCI, the first desktop computer to generate logarithms at a single keystroke, which formed the basis of the later Wang electronic desk calculator. In 1951, he formed Wang Laboratories, Inc., which became a major computer giant in the 1970s and 1980s, when Wang's own inventions helped Wang Laboratories become a major manufacturer of the prototypical desktop computers used in laboratories and schools. Throughout those years, Wang oversaw an uninterrupted series of more compact and efficient instruments and systems for use in office automation and information processing. Wang was also a noteworthy philanthropist.[12]

THOMAS J. WATSON, JNR (1914–1993) – COMPUTER PIONEER AND CHIEF EXECUTIVE

One the most noted chief executives of the twentieth century, Thomas J. Watson, Jnr. led IBM – as chairman and CEO – from the age of mechanical tabulators and typewriters into the computer era. In the process, he brought enormous IT advancements to the world at large. During his leadership, IBM grew from a medium-sized business to one of the dozen largest industrial corporations in the world.

Son of Thomas J. Watson, IBM's first CEO, the younger Watson fought his father's iron-clad will to push IBM into the computer field. As the demand grew for high-volume information processing, the younger Watson's views prevailed. Watson also broke with his father in establishing a more relaxed, decentralized style of management.

Among his first accomplishments upon taking over the company was a major realignment, splitting IBM into six autonomous divisions and the World Trade Corp. Mr Watson stepped down as chairman and CEO in 1971, a year after suffering a heart attack. He remained a member of IBM's board of directors until 1984, taking time out from 1979 to 1981 to serve as US Ambassador to the Soviet Union. He died in Greenwich, Connecticut, on December 31, 1993 at the age of 79.[13]

STEVEN WOZNIAK – COMPUTER PIONEER AND ENTREPRENEUR

The lesser known partner in the Apple computer empire, Steven Wozniak – or "the Woz" – founded Apple Computers Inc. with Steve Jobs in 1976 (see above). Wozniak was the technical wizard behind many of the Apple products and is known as the inventor of Apple II. Experts credit Wozniak with the control program that helped launch the Apple II. During his time with Apple, Wozniak helped write some math routines for a spreadsheet product that Apple had planned to release in competition with VisiCalc and he also worked on the Macintosh. A plane crash affected his memory during the mid-1980s and he never replicated his early successes. In 1985 Jobs and Wozniak received the National Technology Medal from President Reagan at the White House and soon after Wozniak left Apple. Wozniak has all but disappeared from public view.[14]

GLOSSARY – BASIC COMPUTING/WEB TERMS

Bytes, kilobytes, megabytes and gigabytes – The byte is the basic unit of measurement in a computer. Asking what a byte is is about as useful as asking what a second is. A kilobyte is 1024 bytes. And a megabyte is 1024 kilobytes. A gigabyte is 1024 megabytes.

Hard disk and RAM – Because both the hard disk and RAM use the same system of measurement, people often conflate the two. For example, 256 megabytes of RAM is a lot, but these days 256 megabytes on your hard disk is not. So what's the difference? RAM is a Three Letter Acronym (TLA) that stands for Random Access Memory. When you load programs and your personal files, the operating system takes them off the hard disk and puts them into RAM.

Think of your hard disk as an airport and your computer's RAM as the sky. In this analogy the word processor on your hard disk is like an airplane. Your files are like the airplane's passengers. Until the plane loads its passengers and takes off, it sits on the runway, or hard disk.

Operating system - A computer operating system sits between the computer and its peripherals (the hardware) and the programs (word processors, spreadsheets, browsers) you run. It is your computer's underwear, but not the lacy kind. The operating system is practical, in that it provides services for the programs that run on it. For instance, when you save a file in your word processor, the word processor doesn't really do much of the work. Instead it supplies the operating system with information about the file and asks it to save the file. Because the operating system knows where it can put the file on the hard drive, the word processor doesn't have to perform that function.

Clients and servers - You will see or hear the paired terms client/server in almost every discussion of computer networks. Client/server computing is a very simple concept. A client is a program that requests services from another program called a server. It's very much the same as when you, the client, ask a waiter, the server, to bring you a meal. In computing, however, clients and servers run, more often than not, on different computers. Many of the programs you use to do things on the Internet are client programs that talk to server programs on the other computers all over the world.

Browser - Your Web browser is an example of a client program. It asks Web servers anywhere on the Internet to send you the Web pages you request. When you click on one of your bookmarks to go to a Website, the browser asks the Web server at the address of the site to deliver the Web page.

HTTP - Hypertext Transfer Protocol is the language that allows Web servers and browsers to communicate.

URLs - The Universal Resource Locator (URL) is the part of the HTTP protocol that allows browsers to locate servers. URLs have become synonymous with Web addresses.[15]

NOTES

1 Source is Ibiblio Website (ibiblio.com)
2 Source is Jones Digital Century's online encyclopedia.
3 Source is Microsoft Encarta 97 Encyclopedia, 1993–1996, © Microsoft Corporation.
4 Source is MyPrimeTime Website.
5 Source is the Bootstrap Institute Website.
6 Source is the Microsoft Corporation Website.
7 Source is Yale Department of Computer Science Website.
8 Source is the Virginia Tech Website.
9 Adapted from bios on the Texas Instruments and MIT Websites.
10 Source is the UCLA Website.
11 Source is the University of Regina Website.
12 Source is the MIT Website.
13 Source is the IBM Website.
14 Source is Virginia Tech Website.
15 Zuckerman, A. (2001) *Tech Trending*. Capstone, Oxford, pp.83–5. Source is Seth Rothberg, principal, Home-Industries, Amherst, Massachusetts.

Resources

There are almost too many resources to choose from that relate to managing technology. This chapter offers a sampling of some of the best and most informed. It includes:

» key technology trade associations and organizations;
» key books and publications on technology topics; and
» key American government agencies that provide technology assistance.

Thanks to all the advancements in Information Technology outlined in this book, you tech-savvy managers have a World-Wide-Webful of resources to help you cope with the joys and headaches that technology brings you. Figuring out what resources matter and how to locate them is the trial of today's world. Although this section is hardly encyclopedic, it will provide you some of the key places – from associations to books and journals – you can turn to keep up on everything from technology trends to the latest in management practices.

TECHNOLOGY-RELATED ASSOCIATIONS[1]

» **AeA (formerly: American Electronics Association):** 601 Pennsylvania Avenue, N.W.; North Building, Suite 600; Washington, DC 20004; phone – (202) 682 9110; www.aeanet.org.

AeA is reportedly America's largest high-tech trade association. Founded in 1943, AeA has more than 3500 member companies that span the high-technology spectrum, from software, semiconductors, and computers to Internet technology, advanced electronics, and telecommunications systems and services. With 17 regional US councils and international offices in Brussels and Beijing, AeA offers unique global policy grassroots capability and a wide portfolio of valuable business services, products, and events for the high-tech industry, including conferences and publications (see below).

» **ATIS (Alliance for Telecommunications Industry Solutions:** 1200 G Street, NW Suite 500, Washington, DC 20005; phone – (202) 628 6380; fax – (202) 393 5453; www.atis.org.

ATIS was established at the divestiture of the Bell System in 1984. As industry competition grew and new technologies developed, the role of ATIS expanded so that today, ATIS is one of the world's leading standards development bodies for telecommunications, and the leading standards organization for telecoms in North America. ATIS member companies are North American and World Zone 1 Caribbean providers of telecommunications services, and include telecommunications service providers, competitive local carriers, cellular carriers, interexchange companies, local exchange companies, manufacturers, software developers, resellers, enhanced service providers, and providers of operations support.

» **CEA (Consumer Electronics Association):** 2500 Wilson Blvd.; Arlington, VA. 22201-3834; phone - (703) 907 7600; fax - (703) 907 7675; www.ce.org.

Consumer Electronics Association (CEA) membership unites more than 650 companies within the US consumer technology industry. Members are offered exclusive information and top market research, networking opportunities with business advocates and leaders, up-to-date educational programs and technical training, exposure in extensive promotional programs, and representation from the voice of the industry.

» **ECA (The Electronic Components, Assemblies, & Materials Association):** 2500 Wilson Boulevard, Arlington, VA. 22201-3834; phone - (703) 907 7500; fax - (703) 907 7549; www.ec-central.org.

The Electronic Components, Assemblies, & Materials Association (ECA) represents the electronics industry sector, comprising of manufacturers and suppliers of passive and active electronic components, component arrays and assemblies, and commercial and industrial electronic equipment and supplies. ECA provides companies with a link into a network of programs and activities offering business and technical information; market research, trends and analysis; access to industry and government leaders; technical and educational training; and more.

» **EIA (Electronic Industries Alliance):** 2500 Wilson Boulevard; Arlington, VA. 22201-3834; www.eia.org.

The Electronic Industries Alliance (EIA) is a federation of associations and sectors operating in the most competitive yet innovative industry in existence. They are the critical players in their industries. Each has their own members, their own mission, their own autonomy. United under EIA, they form one of the premier high-technology organizations in the world. EIA connects them, not only offering digital-age companies membership in leading associations - but also providing the unique benefits of a true cross-sector alliance.

» **GEIA (Government Electronics and Information Technology Association):** 2500 Wilson Boulevard, Arlington, VA. 22201-3834; general information - (703) 907 7566; membership - (703) 907 7569; fax - (703) 907 7968; www.geia.org.

GEIA represents the high-tech industry doing business with government. Members are companies which provide the government with electronics and information technology (IT) solutions. In partnership with member companies and their government customers, GEIA studies the market for IT, enabling technologies, and advanced electronics products and services for defense and civil government agencies. The organization produces annually a ten-year Defense Electronics Forecast, a five-year Federal IT Forecast, and an Enabling Technologies Forecast, as well as special market studies such as Services & Support, Information Assurance, etc. This is the government outreach arm of EIA.

» **The JEDEC Solid State Technology Association:** (formerly the Joint Electronic Device Engineering Council) 2500 Wilson Blvd; Arlington VA. 22201–3834; www.jedec.org.

JEDEC is the semiconductor engineering standardization body of the Electronic Industries Alliance (EIA), a trade association that represents all areas of the electronics industry. JEDEC conducts its work through its 48 committees/subcommittees that are overseen by the JEDEC Board of Directors. Presently there are about 300 member companies in JEDEC, including both manufacturers and users of semiconductor components and others allied to the field.

» **Information Technology Industry Council (ITI):** 1250 Eye Street NW Suite 200; Washington, DC. 20005; phone–(202) 737 8888; fax–(202) 638 4922; www.itic.org.

The Information Technology Industry Council (ITI) represents the leading US providers of information technology products and services. In 2000, ITI member companies employed more than 1mn people in the United States and exceeded $668 bn in worldwide revenues.

ITI also sponsors the National Committee for Information Technology Standards (NCITS), which was formerly the Accredited Standards Committee X3, Information Technology.

» **NCITS:** (pronounced "Insights") develops national standards and its technical experts participate on behalf of the United States in the international standards activities of ISO/IEC JTC 1, Information Technology. Through participation in NCITS, industry leaders and users alike have the opportunity to open new markets, dismantle non-tariff trade barriers, and build the basic structure of the global information infrastructure.

» **SCTE (Society of Cable Telecommunications Engineers Inc.):** 140 Philips Road, Exton, PA. 19341; phone – (610) 363 6888 or (800) 542 5040; fax – (610) 363 5898; www.scte.org.

SCTE is a nonprofit professional association dedicated to advancing the careers and serving the industry of telecommunications professionals by providing technical training, certification and standards. Since 1969, SCTE has continually expanded its resources and services to meet the changing needs of its members in a rapidly evolving industry. Today, more than 17,500 engineers, technical professionals, installers, and managers depend upon SCTE to deliver the tools they need to maintain their competitive edge. As the only cable telecommunications organization accredited by the American National Standards Institute (ANSI) to develop technical standards, SCTE provides a neutral forum for professionals to collaborate on standards that lead the way to global compatibility.

» **TIA (Telecommunications Industry Association):** 2500 Wilson Blvd., Suite 300, Arlington, VA. 22201–3834; phone – (703) 907 7700; fax (703) 907 7727; www.tiaonline.org.

TIA is one of the leading trade associations in the communications and information technology industry, with proven strengths in market development, trade promotion, trade shows, domestic and international advocacy, standards development, and enabling e-business. Through its worldwide activities, the association facilitates business development opportunities and a competitive market environment. TIA provides a market-focused forum for its more than 1,100 member companies that manufacture or supply the products and services used in global communications. TIA represents the communications sector of EIA.

» **SAE (The Society of Automotive Engineers):** World Headquarters; 400 Commonwealth Drive; Warrendale, PA. 15096–0001; phone – (724) 776 4841; fax – (724) 776 5760.

The Society of Automotive Engineers (SAE) is a one-stop resource for technical information and expertise used in designing, building, maintaining, and operating self-propelled vehicles for use on land or sea, in air or space. The organization is made up of nearly 80,000 engineers, business executives, educators, and students from more than 97 countries, who share information and exchange ideas for advancing the engineering of mobility systems.

ASSOCIATION PUBLICATIONS (PRINT AND ONLINE)

ATIS News – This quarterly publication highlights the activities of all ATIS committees and forums. Also highlighted are new ATIS member companies, new ATIS services, standards, and document releases, and updates on ATIS' key activities, to include testing efforts and industry studies.

ATIS e-Report – This bi-weekly html publication is sent to ATIS members to appraise them of the latest meetings, events, and decisions made within ATIS' committees and forums. The e-Report is an important vehicle in keeping member companies abreast of the latest news within ATIS and the industry. The e-Report is viewable to ATIS members only.

Cyberstates 2001 – A state-by-state overview of the high-technology industry. This report provides you with new 2000 national and state data on high-tech employment, exports, venture capital investments, and home computer and Internet use. Cyberstates 2001 also includes the latest data on wages, establishments, payroll, and research and development expenditures. The report provides trends for each of these indicators since 1994. Cost: $95 AeA members; $190 non-members.

The Information Technology Industry Data Book, 1960–2010 – compiles statistics on computers and related equipment, software and services, telecommunications equipment and services, business equipment, and manifold business forms. The publication also includes employment and other economic factors in the information technology industry that contribute to the US economy. Other chapters focus on the world economy and geographic distribution of the information technology industry markets, and forecasts of economic trends and domestic demand – providing vital information necessary to broaden current markets, target new markets, and gain an industry overview to increase sales. Cost: $225.00.

TIA ONLINE – This is positioned as a community center for the telecom equipment industry. The Buyer's Guide provides an online source to TIA member companies. Featuring vast International Affairs resources, Standards and Technology data and a variety of telecom information, including other related sites of interest, TIA ONLINE is sure to be more than a one-time visit.

PulseOnline – TIA's monthly online newsletter includes a section devoted to "Newly Published Documents," which provides new project numbers, standards proposals, and new documents available; a regular "Industry Trends" column covering cutting-edge communications technologies and business management trends; new member listings; domestic and international market forecasts; an industry calendar; and a variety of feature stories that highlight association activities and industry issues. Free to members and non-members.

Industry Beat – TIA's weekly e-mail news bulletin is a timely and succinct snapshot of pertinent events, association activities, upcoming exhibitions, and other notices important to the communications industry. Free to members.

Standards & Technology Annual Report (STAR) – The Standards & Technology Annual Report, commonly referred to as the "STAR," highlights TIA's engineering activities and the important role of standards setting for the industry. TIA is an American National Standards Institute-accredited standards development agency. STAR features highlights of all TIA engineering committee activities, as well as feature articles on standards-related topics. Free to members and non-members.

Public Policy Report and Agenda – The Public Policy Report and Agenda covers the regulatory issues impacting communications equipment manufacturers. The publication details TIA's policy positions on issues ranging from spectrum management to global trade. A series of charts and graphs serve as benchmarks for competition in the telecommunications industry.

TIA Directory Desk Reference – This publication is a highly visible reference source for the LEC, CLEC, IXC, wireless carrier, ISP, and CATV markets. It is designed to accentuate the capabilities and technological innovations of TIA and MMTA member companies. The directory features company profiles, a buyer's guide of member products and services, and information on the industry sectors and markets that TIA members serve.

2001 MultiMedia Telecommunications Market Review & Forecast – This provides timely access to industry intelligence that can dramatically affect business development strategies. (TIA publication.)

Global Wireless Standards Guide – This was used as the official guidebook for the Informational Session on Global Wireless Standards, a conference held at the 1998 International Telecommunication Union Plenipotentiary.

2001 TIA Directory and Desk Reference – The 2001 TIA Directory and Desk Reference features the equipment, products, and services offered by TIA's member companies. Published annually, the directory provides an encyclopedia of telecom industry information. The directory sells for $129.00 to members and $199.00 to nonmembers.

THE JOURNALS, MAGAZINES AND NEWSPAPERS THAT ANALYSTS TOUT

Although analysts will tell you they read the trades and mainstream media "with a grain of salt," read they do. Of late, the Web is becoming a prime source of information for the experts.

Besides the fat glossy mags like *Fast Company* and *Business 2.0*, analysts point to lesser known journals such as *The Industry Standards*, which is all about Internet development and appears both in print and in on the Web. *Transportation & Distribution* won't be known to many of you outside the transport industry, but some analysts prefer its high-tech coverage to just about anything else. And they say the same for *Supply Chain Technology News*, and *Technology Review: MIT's Magazine of Innovation*.

Here are the general news and business publications that the analysts are reading – *Success, Forbes, Fortune, Business Week, Red Herring, Economist, The Wall Street Journal, The New York Times, USA Today, The Washington Post, Harvard Business Review, Time, Newsweek*.

The high-tech industry publications most touted are - *Information Week, Computerworld, InfoWorld, Network World, Business 2.0, Fast Company, Wired, IndustryStandard, PC Magazine, eWeek* (formerly *PC Week*), *Interactive Week, CIO* magazine, *Knowledge Management Review* (KM), *Complexity Journal*.

And on the Web they say check out - msnbc.com (technology), Yahoo.com, CNET.com, ZDNET.com, LocalBusiness.com, and Upside.com.[2]

COMMERCIAL IT PUBLICATIONS[3]

Zuckerman, A. (2001) *Tech Trending: A Visionary Guide to controlling your Technology Future* (foreword by Stewart Chiefet). Capstone, Oxford.

Labovitz, G. & Rosansky, V. (1997) *The Power of Alignment: How Great Companies Stay Centered and Accomplish Extraordinary Things*, 1st edn. John Wiley & Sons, New York.

Spewak, S.H. & Hill, S.C. (1993) *Enterprise Architecture Planning: Developing a Blueprint for Data, Applications and Technology*. John Wiley & Sons, New York.

Davenport, T.H. & Beck, J.C. (2001) *The Attention Economy: Understanding the New Currency of Business*. Harvard Business School Press, Boston.

Peltier, T.P. (2001) *Information Security Risk Analysis*. Auerbach Publications, New York.

Shapiro, H. & Varian, H.R. (1998) *Information Rules: A Strategic Guide to the Network Economy*. Harvard Business School Press, Boston.

Groth, D., Newland, D., *et al.* (2001) *A+ Complete Study Guide*, 2nd edn. Sybex, San Francisco.

Gibas, C., Jambeck, P. (2001) *Developing Bioinformatics Computer Skills*, 1st edn. O'Reilly & Associates, Cambridge, MA.

Duda, R.O., Hart, P.E. & Stork, D.G. (2000) *Pattern Classification*. John Wiley & Sons, New York.

Reddy, M.T. (1995) *Securities Operations: A Guide to Operations and Information Systems in the Securities Industry*. Prentice Hall Press, New York.

Linthicum, D.S. (1999) *Enterprise Application Integration*. (Addison-Wesley Information Technology Series) Addison-Wesley Publishing Company, Reading, MA.

Cover, T.M. & Thomas, J.A. (1991) *Elements of Information Theory*. John Wiley & Sons, New York.

Spofford, M. (2001) *MDX Solutions: With Microsoft SQL Server Analysis Services*, bk & CD-ROM edn. John Wiley & Sons, New York.

Collins, H. (2000) *Corporate Portals: Revolutionizing Information Access to Increase Productivity and Drive the Bottom Line*. AMACOM, New York.

Sturm, R., Morris, W. & Jander, M. (2000) *Foundations of Service Level Management*. Sams, Indianapolis.

Coe, M. (1996) *Human Factors for Technical Communicators*, 1st edn. John Wiley & Sons, New York.

Erik Thomsen, E. (1997) *OLAP Solutions: Building Multidimensional Information Systems*, bk & CD-ROM edn. John Wiley & Sons, New York.

Hall, S.H., McCall, J.A., Hall, G.W. & Hall, S.E. (2000) *High-Speed Digital System Design: A Handbook of Interconnect Theory and Design Practices*. John Wiley & Sons, New York.

Kanter, R.M. (2001) *Evolve!: Succeeding in the Digital Culture of Tomorrow*. Harvard Business School Press, Boston.

Shuman, J.C., Twombly, J. & Rottenberg, D. (contributor) (2001) *Collaborative Communities: Partnering for Profit in the Networked Economy*. Dearborn Trade, Chicago.

Oram, A. (ed.) (2001) *Peer-to-Peer: Harnessing the Power of Disruptive Technologies*. O'Reilly & Associates, Cambridge, MA.

Robinson, M., Tapscott, D. & Kalakota, R. *e-Business 2.0: Roadmap for Success* (Addison-Wesley Information Technology Series), 2nd edn (ed. M. O'Brien). Addison-Wesley Publishing Company, Reading, MA.

Brown, J. (2000) *Minds, Machines and the Multiverse: The Quest for the Quantum Computer*. Simon & Schuster, New York.

Pipkin, D.L. (2000) *Information Security*. Prentice Hall PTR, NJ.

Kurzweil, R. (2000) *The Age of Spiritual Machines: When Computers Exceed Human Intelligence*. Penguin USA.

Pottruck, D.S. (2000) *Clicks and Mortar*. Jossey-Bass, San Francisco.

Cassidy, A. (1998) *A Practical Guide to Information Systems Strategic Planning*. CRC Press – St. Lucie Press, Delray Beach, FL.

Britton, C. (2000) *IT Architectures and Middleware: Strategies for Building Large, Integrated Systems*. Addison-Wesley Publishing Company, Reading, MA.

Moore, G.A. (2000) *Living on the Fault Line: Managing for Shareholder Value in the Age of the Internet*. HarperBusiness, New York.

Laudon, K.C. & Price Laudon, J. (1999) *Management Information Systems: Organization and Technology in the Networked Enterprise*. Prentice Hall, Englewood Cliffs, NJ.

Allen, P. (2000) *Realizing eBusiness with Components*. Addison Wesley Professional, Reading, MA.

Ward, J. & Griffiths, P.M. (1996) *Strategic Planning for Information Systems* (Wiley Information Systems Series). John Wiley & Sons, New York.

US GOVERNMENT E-COMMERCE INFORMATION SOURCES

The following US National Institute of Standards and Technology (NIST) departments focus on e-commerce and can help global managers get started with useful information and even advice:

» *ATP Information Technology and Applications Office*: The ATP fosters partnerships among government, industry and academia. It co-funds high-risk, industry-proposed research to develop enabling technologies that promise significant commercial payoffs and broad economic benefits for the nation. The ATP provides a mechanism for industry to extend its technological reach and push out the envelope of what can be attempted. Its Website is http://www.atp.nist.gov/itao/

» *Systems Integration for Manufacturing Applications (SIMA)*: The SIMA program seeks to improve the overall integration of manufacturing systems by facilitating development and test of information exchange protocol standards, and to foster industry awareness of advances in integration technology. SIMA works with manufacturers, software vendors, standards organizations, other government agencies, and researchers to achieve these objectives. Its Website is http://www.mel.nist.gov/msid/sima/sima.htm

» *eBusiness Product Lines*: The manager of this small department is responsible for the product line strategy and the development and deployment of the electronic business and electronic commerce products and services through a nationwide system of manufacturing consulting offices. The Website is http://www.mep.nist.gov

» You can locate the Computer Security Resource Center at http://csrc.nist.gov[4]

ONLINE STANDARDS INFORMATION

All major international standards organizations offer a wealth of information on their Websites. Just take the acronyms – ISO, IEC or ITU – and tag the www. prefix on the front and .org suffix on the back and you'll have instant access. The following is on the American National Standards Institute's (ANSI) online services, almost all of which offer international outreach. ANSI has a number of online services to help global managers keep tabs on standards and testing practices. At www.ansi.org look for the following.

Deposit accounts

The ANSI Online Electronic Standards Store (ESS) has added a new Deposit Account feature enabling customers to purchase individual copies of electronic documents at any time, without requiring the use of a credit card to complete the transaction.

The ESS is an Internet-based e-commerce site that allows customers to:

» search for standards using keywords;
» select documents for delivery in Adobe Acrobat .pdf format;
» purchase selected documents online via a secure server; and
» download the documents directly to their desktop.

For more information on the Deposit Account function, please visit ANSI Online at http://webstore.ansi.org/ansidocstore/dep_acc.asp

The National Standards System Network (NSSN)

This contains a database on more than 12,000 standards – everything from information imaging to storage and Web development – and acts as a national resource for global standards. You can use it to track changes, either existing or upcoming, in any technology you may want to implement.

STAR, the fee-based component of the NSSN, provides an integrated network of databases containing up-to-date information on more than 280,000 approved standards published by the world's leading standards developers. Further, STAR is the only e-mail-based information service known to notify its users when standards are revised or when new development projects are initiated.[5]

NOTES

1 Information is culled from association Websites. Addresses are listed in the text.
2 Zuckerman, A. (2001) *Tech Trending*. Capstone, Oxford, pp.151–2. Sources are Merv Adrian, VP & research manager, Giga Information Group, Cambridge, Massachusetts, and David Yokelson, the META Group, Stamford, Connecticut. Research conducted by Eddy Goldberg and Randy Bernard.
3 Source is Amazon.com and Capstone.
4 Zuckerman, A. (2001) *Tech Trending*. Capstone, Oxford, pp. 169–70.
5 *Ibid.*, pp.172–3.

Ten Steps to Making Managing Technology Work

Learning about managing technology is one thing; making it work for you is another. This chapter looks at a number of the key elements raised throughout this guide and offers tips and advice on how to apply them in the field. It includes:

» an emphasis on training and organizational skills;
» the need to select technology from a business strategy perspective; and
» tips on keeping your sense of humor throughout this process.

Managing an organization of any kind in a high-tech world is as much an art as a science. Yes, you will need rudimentary knowledge of technology trends that will affect your operation in the years to come and you need training in the skills required to handle that technology effectively. But just as crucial is having a feel for the people in your organization, their emotional timbre and ability to cope with the change that technology ultimately brings with it. And keeping a sense of humor may be the most desirable asset to cultivate.

The following are ten steps that every tech-savvy manager should follow to keep abreast of the technology wave; to ensure that you are using technology wisely and strategically, and that you are driving the technology change in your organization rather than allowing it to drive you.

1. DON'T SKIMP ON TRAINING

We have witnessed enormous technology change since the end of World War II, and it seems that change is accelerating each year. The experts call that Moore's Law. You have to be prepared to make technology learning a part of your job and every employee's work week. And that's true for the CEO as well as the starting employee.

As noted in Chapter 2, managers have to know the basics of word processing and have some passing understanding of the technology that already exists in their organization. Let your IT department and vendors teach you the basics and then make sure that you book into your week the time to learn about technology developments that could affect your organization. Remember, you're not aiming to become a techie – just to be tech-literate, tech-aware, and tech-savvy.

2. READING, WRITING, AND ANALYSIS MATTER

You can't be tech-literate, tech-aware, and tech-savvy if you can't read for comprehension, analyze, and write a decent sentence. All that enormous amount of information you are collecting in all those servers requires different skills than those taught in Computing 101. You and your employees could benefit from some of the training that journalists receive, particularly an emphasis on accuracy. As one executive of a major credit card company has pointed out, it used to be one error affected 5000 customers. Today, with Internet transmission, the same error can hit five million people in a matter of minutes. Errors mean lost time and lost profits.

3. KNOWING THE TRENDS

As noted in Chapter 2, tech-savvy managers must be able to track technology trends with an eye to what is here to stay. Without this knowledge you can't make cost-effective technology decisions, organize your company or institution to roll with technology change, or train yourself and your employees in the skills they'll need to survive in the New Economy.

But just knowing about any technology won't make you a strategic IT thinker and manager. Best to hone your attention to four major areas – communication, which includes the Web and wireless technologies; e-commerce and business-to-business e-commerce (B2B); the move from a supply chain to a supply network; and technologies affecting transportation and logistics.

4. BEATING THE ANALYSTS

You don't want to be reliant on analysts, consultants, and vendors to call the shots for you, alone. To better assess their advice, here are some tricks of the analyst's trade.

» *Filtering information:* The first and foremost practice analysts provide is filtering information, as in sorting out the significant from the insignificant. They also keep an eye on emerging trends that may not appear important at the time, but could be the next big buzz. Analysts themselves warn that we should beware of their bias towards the Fortune 1000 companies and technology with some track record. Yes, track records and experience are important in selecting technology (See Chapter 7), but tech-savvy managers should at least have antennae out for what's new or untried.

» *Good note-taking, research, and classification skills:* If you want to play the analyst's game be sure to hone those listening skills so you can take good notes. You have to want to conduct sometimes mind-numbing amounts of research and be able to classify – or break out – information to create the categories that will indicate trends or other information worth analyzing.

» *Be able to make relativistic comparisons:* Oftentimes there are no statistics or absolute benchmarks to serve as the basis for analysis. Analysts make what they call "relativistic comparisons" to offer perspective for clients. One way to make those sorts of comparisons is to learn from your clients and recognize that every interview offers a two-way exchange and educational opportunity.

» *Turn experiences into case studies:* People always learn more easily when you either present a visual or tell a story. Analysts learn quickly to turn their experience into easily digestible case studies. Case studies also turn them into instant experts and allow them to infer vast knowledge from shallow samples. You can do the same when making a case to your organization for the purchase of a solution, for example.

» *Be able to map old problems onto new trends:* Grasping where yesterday's problems match today's world, or yesterday's solutions fit today's market, is a big analyst trick. Sometimes this involves being able to benchmark – or compare best practices, company to company – and sometimes it means simply swapping names. So a service bureau now becomes an ASP because this is a hot market.[1]

5. PLAYING THE STANDARDS/CONSORTIA GAME

The world of standards and high tech consortia can seem arcane, at first, but all the really tech-savvy managers know that many decisions affecting product design and development are hatched in these organizations. Knowing which working group or committee is working to standardize the technology that will affect your organization will keep you ahead of the pack. The following listings should help you start connecting the dots between technologies, standards organizations, and consortia.

Cellular phones

» Alliance for Telecommunications Industry Solutions (ATIS); Secretariat of ANSI-Accredited Standards Committee T1–Telecommunications (ASC T1)

» Telecommunications Industry Association (TIA)
» Electronic Industries Association (EIA)
» International Telecommunication Union (ITU)

Semiconductors

» Joint Electronic Device Engineering Council (JEDEC)
» Electronic Industries Alliance (EIA)
» Semiconductor Equipment and Materials International (SEMI)
» Institute for Electrical and Electronics Engineers (IEEE)
» American Society for Testing Materials (ASTM)

Internet/Web

» Internet Society (ISOC)
» Internet Engineering Task Force (IETF)
» W3C
» ISO/IEC JTC 1 on Information Technology
» ASTM (related to healthcare transactions only)
» Alliance for Telecommunications Industry Solutions (ATIS); Secretariat of ANSI-Accredited Standards Committee T1–Telecommunications (ASC T1)
» Institute for Electrical and Electronics Engineers (IEEE)
» International Telecommunication Union (ITU)
» Organization for the Advancement of Structured Information Standards

Business-to-business processing

» RosettaNet

Bar codes

» ISO/IEC JTC 1/SC 31
» AIM USA (AIM is the global trade association of providers and users of components, networks, systems, and services that manage the collection and integration of data with information management systems)
» Uniform Code Council (UCC)
» Graphic Communications Association (GCA)
» HIBCC (Health Industry Business Communications Council)

Intelligent Transportation Systems (ITS)

» ITS America
» IDB Forum (For Inter Data Bus)[2]

6. KEEPING TABS ON NEW IT JOBS AND WHAT THEY MEAN

New Economy jobs are being created right and left to fill the need that the Web-enabled world is creating. Chief Knowledge Officers (CKOs), Managers of Information Services, and Chief Information Officers (CIOs) have become familiar to anyone operating a mid-size or larger organization. Most prevalent are CIOs, who generally report to the CEO and oversee a company or institution's IT activities. With the advent of the World Wide Web, CIOs have to be able to tackle new problems in cyberspace. To handle these new requirements CIOs have to be able to:

» tackle and utilize the new Web-based technologies;
» provide rapid delivery of solutions and transfer of best practices across an enterprise;
» function outside the enterprise;
» evaluate technology requirements throughout a supply network, as well as in cyberspace;
» evaluate new Web-based markets;
» develop B2B strategies in conjunction with ongoing business strategies, and tie them in;
» develop e-commerce teams or even divisions;
» realize the impact of e-commerce on the entire organization;
» work with many more people in a collaborative fashion;
» promote accuracy throughout a supply network; and
» cope with an expanded, more visible role.[3]

7. MANAGING CHANGE

Here comes the "art" part – managing technology change. As noted in Chapter 2, technology will make change. That's the most important factor strategic decision-makers, professionals, and entrepreneurs have to know and accept to stay competitive while managing networked/B2B-oriented organizations. Change is confusing, difficult, and often painful. But, as all of you know, there's no way to avoid technology change. Accepting the change technology brings rather than resisting it will help ease you into a world in which technology serves you and your organization, rather than you serving it. And knowing the emotional stages that both you and your employees will experience will make you better equipped to promote change within your organization and reap technology's benefits.[4]

The following are crucial steps you need to take to manage any organization today that relies on advanced technology:

» break down hierarchical structures to allow for information sharing and true collaboration among employees and then throughout a supply network;
» create new management positions to ensure proper use of technology throughout your organization;
» improve communication skills so that management clearly expresses priorities;
» build customer relationships rather than only providing customer service; and
» establish new rules of communication so that the vast amount of information that technology makes available can be used for maximum purpose, and not overwhelm an organization.[5]

MANAGING EMPLOYEE ANGER AND NEGATIVITY

The British Standards Institution (BSI) has decided to institutionalize the process of managing the employee anger and negativity that technology can create, through creation of its guide BS 8600 for Complaints Handling. It may seem rather silly to codify a process that amounts to allowing employees to mouth off, but smart managers have long understood that giving employees a voice in a company boosts morale and opens communication channels that can lead to important improvements.

BSI officials report that their new guide is about how an organization might put in place a system for handling complaints. It's for anyone, large or small, and for all divisions, not just manufacturing.

BS 8600 leaves it up to the organization to decide what sort of system they want for handling complaints. As for how many companies will want to be this systematic about allowing their employees to vent their feelings, no one knows quite yet. It will take a year or so to see if this approach works any better than threatening to fire employees if they don't endorse that new ERP system.[6]

8. ORGANIZING FOR A NETWORKED, B2B FUTURE

As always, business strategy should be top of the list when dictating a technology course. Management issues are also key. You will want to explore how information flows in your company, and the sorts of skills your employees have or lack, to handle the tasks of working in an networked, Web-based fashion. Ensuring that employee schedules are flexible to meet the demands of this new world is also crucial.

As noted in Chapter 2, tech-savvy managers know that these are elements that relate to management and organization. If you take the time up-front to establish human procedures and rules, processes and communication channels, you will find the job of managing in a networked, Web-enabled world far easier and you will more than earn back the cost of technology implants. Knowing how to organize your operation to function in both three-dimensional and virtual worlds, promote communication and accuracy, and operate electronic equipment and computing is key to success.

9. TECHNOLOGY CHOICES THAT SUPPORT YOUR STRATEGY

Every manager at all levels of an organization may be confronted with making technology choices. The key to a wise, cost-effective approach to this process is knowing and sticking to business priorities while keeping an eye out for technology that will promote goals and be flexible enough to grow with you and scalable enough to be compatible with future technology developments.

Although this was pointed out in Chapter 2, it bears repeating: learning how to conduct IT assessments is crucial to your success.

Basically, an IT assessment is a method for both cataloging technology and comparing what you have in-house with the benefits or deficits it produces. This information is used as a benchmark device against future technology purchase. Managers starting from scratch need to spend some quality time researching what's on the market.

10. KEEPING A SENSE OF HUMOR

Who really loves technology? Who wants to cuddle up to a monitor or caress a keyboard? Not many of us. But tech-savvy managers appreciate what technology does for their organization. And the smartest of them learn to laugh with the constant crashes and upgrades that come with the advanced technology turf.

Believe it or not, there are people who specialize in managing your workplace with a good dollop of humor. Here's what one humor management consultant has to say on the subject: "Humor in the workplace works like a thermostat, controlling the climate within the environment. It is a key component of the ambiance that greets and surrounds everyone who enters the workplace. Simply put, positive humor fosters a warm and inviting feeling. Negative, divisive humor makes a place seem extremely cold and aloof."[7]

If you can face technology change with a smile on your face you'll know that you may well be on the way to developing that networked, B2B organization that you've outlined as key to your business and institutional success. And if you can't quite manage a smile, try writing about your technology frustrations to friends and colleagues. E-mail is a good technology tool for downloading your frustrations. At the very least this approach keeps you from firing the IT department, smashing your terminal or shredding that report.[8]

KEY LEARNING POINTS

» Don't skimp on training for yourself or your employees. Expect tech training to be a life-long experience.
» Reading, writing, and analysis are crucial to managing data and information and key to competitive advantage.

» Knowing what tech trends are here to stay will help you prepare your organization to function in a networked, Web-enabled world.

» Don't be reliant on the information that analysts, consultants, and vendors provide you. Learn their skills, bone up on tech topics you are discussing with them, and be prepared to question what they have to say.

» Playing the standards and consortia game is a major factor in tracking tech trends before they go public and keeping your organization competitive.

» Tech-savvy managers need to keep tabs on evolving jobs in the IT arena and what they can mean for your organization.

» Knowing how to manage the change that technology inevitably brings will be a core part of your job, and will mean success in the long run.

» Don't plunk technology into your operation without organizing for proper information flow and communication.

» Make sure the technology you select supports your core business or institutional strategy.

» And keep a sense of humor about all the annoyances and foibles that incompatible technologies bring to the job.

NOTES

1 Zuckerman, A. (2001) *Tech Trending*. Capstone, Oxford, pp.156–7.

2 *Ibid.*, pp.163–4. Source is Stacy Leistner, director of communications and public relations, ANSI.

3 *Ibid.*, p.223.

4 *Ibid.*, p.191.

5 *Ibid.* p.193. Source is David Washburn, Principal Consultant at Amherst Information Architects, Amherst.

6 *Ibid.*, p.211. Source is John Hele, BSI business manager for quality systems.

7 *Ibid.*, p.224. Source is Gesell, I. *HUMOR: The Universal Human Resource*.

8 *Ibid.*, p.225.

Frequently Asked Questions (FAQs)

Q1: Why is technology training so important?

A: See Chapter 1 and Chapter 10, Section 1.

Q2: Why do human skills like reading and analysis matter?

A: See Chapter 1 and Chapter 10, Section 2.

Q3: What technology trends matter most to my organization?

A: See Chapter 1, Chapter 2, and Chapter 3.

Q4: How has technology affected supply chain development?

A: See Chapter 3, Section "How technologies are evolving the supply chain into a supply network," and Chapter 4.

Q5: What problems emerge when I practice e-commerce?

A: See Chapter 5.

Q6: How is the world of procurement and sourcing changing?

A: See Chapter 3, Chapter 4, and Chapter 5.

Q7: What technology will I need to move to a net-worked, Web-enabled world?

A: See Chapter 4.

Q8: Who are the key thinkers and entrepreneurs whose work has affected my operation?

A: See Chapter 8.

Q9: What resources are available to me to help me cope with technology?

A: See Chapter 9.

Q10: What are some best practices available from successful supply chain implementations?

A: See Chapter 7.

Acknowledgments

Thanks to David Washburn, principal of Amherst Information Architects, who provide IT background and expertise; and to Eric Goldscheider, Eddy Goldberg, Randy Bernard and Rosalind McLymont for their top-notch research. Kevin Fitzgerald, editor-in-chief and associate publisher of *Supply Strategy* and Jean Murphy, publisher, *Global Sites and Logistics*, provided information and content from their publications. And Don Moon of the Wesleyan University Government Department deserves thanks for offering a political perspective.

Index